AF746832
Bernard King Library
318101
WITHDRAWN

for Global Industries

Oil and Gas

A study and practice book for oil and gas professionals

Steve Oliver

Garnet EDUCATION

Published by
Garnet Publishing Ltd.
8 Southern Court
South Street
Reading RG1 4QS, UK

Copyright © Garnet Publishing Ltd 2010

The right of Steve Oliver to be identified as the author of this work has been asserted by him in accordance with the Copyright, Designs and Patents Act 1988.

All rights reserved.

No part of this publication may be reproduced, stored in a retrieval system, or transmitted in any form or by any means, electronic, mechanical, photocopying, recording or otherwise, without the prior permission of the Publisher. Any person who does any unauthorized act in relation to this publication may be liable to criminal prosecution and civil claims for damages.

First published in 2010.

ISBN 978 1 85964 506 2

Production

Project manager:	Sarah Mellowes
Editorial team:	James Croft, Simone Davies, Sarah Mellowes, Rod Webb
Design & Layout:	Christin Helen Auth
Illustration:	Doug Nash
Technical review:	Oil and Gas UK
Photography:	Corbis, Getty Images, Clipart, iStockphoto

Every effort has been made to trace the copyright holders and we apologize in advance for any unintentional omissions. We will be happy to insert the appropriate acknowledgements in any subsequent editions.

The author and publisher would like to thank the following for permission to reproduce copyright material:
Map on page 9 reproduced with kind permission of BP p.l.c.
BP logo on pages 9, 11 and 51 reproduced with kind permission of BP p.l.c.
SPE logo on page 9 reproduced with kind permission of the Society of Petroleum Engineers.
Text and screenshot on page 12 reproduced with kind permission of LCM Oil & Gas Ltd. (www.lcmoil.com)
Top middle image on page 55 copyright SBM Atlantia, Inc. 2010. All rights reserved.
Bottom right image on page 55 reproduced courtesy of DOSECC.
Photograph on page 76 courtesy of the U.S. Geological Survey.
Image on page 85 reproduced with kind permission of Seabed Rig AS (www.seabedrig.com).
Text on pages 102 and 103 reproduced with kind permission of ABB Asea Brown Boveri Ltd (www.abb.com).
Text on page 122 reproduced with kind permission of Simon Marquis (www.oilrigdisasters.co.uk).
Extract 1 on page 126 from www.api.org/aboutoilgas/natgas/index.ctm (February 12, 2010). Reproduced courtesy of the American Petroleum Institute (www.api.org).
Extracts 2–5 on pages 126 and 127 from www.api.org/ehs/performance/explore/moreexplorproduction.ctm (October 9, 2009). Reproduced courtesy of the American Petroleum Institute (www.api.org).
Photo 3 on page 128 reproduced with kind permission of BJ Services Company.
Photo 4 on page 128 reproduced with kind permission of JuneWarren-Nickle's Energy Group.
Photo 5 on page 128 reproduced with kind permission of Philip Rogers.
Image on page 149 reproduced with kind permission of www.sharkbay.com
Text on pages 154 and 155 reproduced with kind permission of Petroleum News.
Image on page 168 reproduced under a GNU Free Documentation License © Xmantis (www.searchanddiscovery.net).
Text on page 176 reproduced with kind permission of BJ Services Company (www.bjservices.com).
Text and image on page 181 reproduced with permission of Cameron Systems Limited (www.c-a-m.com).
Photo b on page 187 reproduced with kind permission of Acock/Anaqua Operating Co., LP.
Text on page 190 reproduced with kind permission of OilVoice. (www.oilvoice.com).
Text on page 191 reproduced with kind permission of The GasGun, Inc. (www.TheGasGun.com).
Text on page 197 © United States Department of Energy.
Image on page 202 reproduced under a GNU Free Documentation License © Rvaron.
Text on page 206 HOUSTON CHRONICLE by Matthew Leising. Copyright 2004 by HOUSTON CHRONICLE PUBLISHING COMPANY. Reproduced with permission of HOUSTON CHRONICLE PUBLISHING COMPANY in the format Textbook via Copyright Clearance Centre.
Extract 2 text on page 207 reproduced with kind permission of Guardian News & Media Ltd.
Image on page 207 reproduced with kind permission of Wolter van der Pers, Director, International Marsac (www.marsac.nl).
Text on page 210 reproduced with kind permission of Intertek (www.intertek.com).

Every effort has been made to ensure the accuracy of the information contained in this course. However, the oil and gas industry changes at a rapid rate, so information can become out of date. It is also a very complex industry and its essential activities are done in different ways from company to company. We apologize if there are inaccuracies, but please keep in mind that it is a language course and not a technical training course. We hope that it does not affect your enjoyment of the course.

Acknowledgements
I would like to thank all those who have helped directly and indirectly in the writing of this material. In particular, I am grateful to colleagues at Accent International Language Consultancy, especially Mark Waistell, who provided valuable suggestions in the early stages and encouraged me in many ways during the project. I am indebted to the editorial staff of Garnet, especially James Croft and Sarah Mellowes, whose patience and eye for detail have been extraordinary, and to Rod Webb who has believed in me and in the ideas I have attempted to put on paper. Finally, I would like to thank the many learners from the oil and gas and other industries, especially from BP, who have been on the receiving end of the material during its conception. Outstanding of these was Sergey Shcherbakov from TPG in Russia, who provided a platform for my early knowledge and research into the world of oil and gas exploration and production.

Printed and bound
in Lebanon by International Press: interpress@int-press.com

Contents

English for Global Industries: Oil and Gas

Introduction

A Guide to Working with *English for Global Industries: Oil and Gas*

The main purpose of this book is to help you to expand your English vocabulary for use at work in the oil and gas industry. The key to successful vocabulary activation is your engagement in and with the learning process. There is a big difference between the English words that you can understand and the words you can use. This book is intended to help you to increase the English vocabulary that you actually use in your job, improving your communication at work.

Read this introductory guide carefully before you begin working your way through the units that are relevant to your work.

Getting started

Each unit is independent, so you can choose the units whose topics and vocabulary are most useful to you. You do not need to work through the book in sequence. The contents pages for each section list the focused readings and summarize the topics, vocabulary areas and aims of each unit. Once you have chosen a unit to study, follow the learning procedure outlined below. Each is self-explanatory and provides all the instruction you need to move forward. On average, each unit will take around an hour to complete.

Expanding your active vocabulary

There are three steps involved in expanding your active vocabulary:

- Vocabulary building – this involves increasing your stock of new words;
- Vocabulary storage – this involves recording and then remembering the new words;
- Vocabulary activation – this involves using the new words.

This book will help you with all three of these.

Vocabulary building

This involves identifying and understanding new words and recognizing which words are immediately relevant for you. Both the focused reading and vocabulary-building activities represent the vocabulary-building stage of the learning process. The focused readings take you straight into the unit topic, open up new vocabulary areas and introduce the vocabulary-building activities that follow. Extracts have been chosen from a number of industry media sources, including magazines and websites, and should be engaged with critically, as if reading them in the original context. Read the extracts as quickly as possible, then answer the comprehension questions or complete the accompanying exercise. These are designed to help you grasp the key points, so do not worry about understanding every word.

The vocabulary-building activities explore in closer detail the vocabulary areas introduced in the readings. Answers are given in the answer key in the back of the book. The degree of your engagement with the answers will determine how you progress in learning the language. Only check your answers once you have completed all the activities. Example answers are often given to help you make sure that you have understood the activity. The glossary pages at the end of every two units will help you understand any unknown words or phrases.

Key tips as you work through this book:

- Look out for useful words and expressions, when you are using this book and when you are at work. List new words that you come across in a notebook to look up later when you are studying.
- Make sure you have a good English dictionary to support your use of the glossaries and word list provided.
- Note word families and derivations, and common patterns in word endings that indicate the type of word or part of speech. Consider, for example: explore/exploration; quick/quickly; geology/geologist, etc.
- Learn words in a context and in partnership with other words as much as possible (for example, 'shift work', 'run smoothly', 'water saturation', etc.). Recognizing collocation – the way words work together – is extremely important for grasping the meaning of a sentence.
- Select vocabulary for active learning according to your needs. You don't need to actively learn every new word you meet; some can be set aside for later. Ask yourself two key questions: Do I already use it? Will I need to use it in the future?

Vocabulary storage

This involves the ongoing task of keeping a record of any new words which you think are useful. The word list at the back of the book is designed as a referencing tool; to help you access the context of first appearance and as a quick reference. The glossaries give new words in the context of phrases and word partnerships. Follow these examples, using full sentences to ensure that there is a clear context for each word. A number of other ways of storing vocabulary are suggested below. We suggest that you alternate your use of these methods and that you review and reorganize your records regularly.

- Use a diagram to display words in a context (see page 55, for example).
- Categorize words in families of words, using a table, for example, with columns to include the different members of the 'family' (see page 10, for example).
- Record individual words on cards and use the cards for 'playing' with words, to help you remember them. You can make cards for opposites, for synonyms, or words with a similar meaning and with pictures, or with written definitions.
- Use colour pens or highlighting to help new words become more accessible and more memorable.
- We recommend that you read about and experiment with 'mind-mapping', a technique used to great effect by Tony Buzan. He has some inventive ideas for vocabulary storage.

Vocabulary consolidation and storage is an ongoing task, but following the conclusion of the vocabulary-building activities, the 'Refresh' box reminds you to look back over the work you have done before going on to the 'Put it to work' activity.

- Read the focused reading extracts again to check or get a better understanding.
- Use the glossary to understand any unknown words.
- Identify any new vocabulary which you think will be useful for you.
- Record it in an effective way, using some of the ideas for storing vocabulary mentioned above.

Vocabulary activation

It is very important to activate new vocabulary as soon as possible after you learn it, by using it in writing or speaking. If you are working in English, or if you are having classes in English, then you will have plenty of opportunities to do this. If you are using this book on your own, then you will need to practise with writing. If you can find a partner to work with, you can do this through e-mail or web-chat.

To activate vocabulary as effectively as possible:

- Experiment as much as possible when you speak and write. Play with words.
- Be prepared to make mistakes. If you are not making mistakes, it may be because you are not learning anything new.
- As you review your vocabulary storage, ask yourself whether you are actually using your new vocabulary in writing or in conversation.
- Set yourself targets to use particular words or expressions in particular situations (in a meeting or on the telephone, for example).

Although vocabulary activation is an ongoing process, the first step in vocabulary activation will be the 'Put it to work' activity. This is designed to help you understand each unit's learning objective, providing an opportunity for you to employ the language you have learnt in a real-world context.

Notes for trainers using *English for Global Industries: Oil and Gas* in a classroom setting

Although designed for self-study, this book is easily adaptable for classroom use. With the addition of oral activities, based on the material in the book, but guided by you, your students will benefit further. On average, each unit will take around 90 minutes of classroom time to complete. We would recommend that the material is used in class in the same way as is suggested on the previous pages.

The following tips will help you to enhance the students' learning experience.

- *Allow students to select which units they will study.* If you are working one-to-one with a learner, then he or she will be able to select the units they want to study. If you are working with a group of students, let the group members decide by consensus. Alternatively, you can make the choice based on your knowledge of their needs and interests. As with all ESP training, needs analysis will be the key to this decision.
- *Encourage independent learning.* By all means highlight and explain new vocabulary to the learners and discuss lexical aspects, but do not let the students become too dependent on you. It is important for your students to become autonomous in their vocabulary learning and they should be encouraged to think for themselves as much as possible.
- *Facilitate by providing answers at appropriate moments.* You will be able to provide the answers at the appropriate moment without them having to look them up in the back of the book. This should help keep a steady pace during learning.
- *Encourage discussion.* In the classroom setting, your students will have the opportunity to discuss activities with you and each other in pairs or groups as appropriate. Encourage discussion of any answers that are felt to be inaccurate. This builds confidence and helps with vocabulary activation and provides a valuable 'rounding-up' and consolidation of the work that has been done.
- *Provide feedback.* Use a variety of means to provide feedback to learners, including written notes, audio or video recording, language clinics, etc.
- *Take advantage of experts in the field.* Try to involve experts from among your class to enhance the activities in the book with their knowledge and experience. You will find that there are extensive supplementary resources, including authentic up-to-date documents, that can be made available by students to enhance learning.
- *Introduce review activities* at suitable points during your course to further consolidate and enhance their learning.

Section 1 Contents

The Business of Oil and Gas

	Titles	Focused Reading	Vocabulary Areas	Aim(s)
1	A Global Giant	• Online encyclopedia introduction to the oil and gas industry • Quotes from oil and gas company websites • Map showing proved oil reserves, end 2007	• Essential oil and gas vocabulary • Essential business vocabulary	To be able to describe the oil and gas industry in a particular country.
2	Presenting the Company	• Company website profile • PowerPoint presentation slide	• Company performance and strengths • Marketing and promotion	To be able to present a company.
3	The Marketing of Oil and Gas Services	• Job advertisement for a Marketing Communications Leader • Transcript of a telephone conversation about marketing	• Products and services • Price • Other marketing language	To be able to describe the products or services sold by a company and how they are marketed.
4	Oil and Gas Accountancy	• Introduction to the complexities of accounting in the oil and gas industry • Feasibility study cost projection	• Accountancy terms • Methods of accounting • Cost analysis	To be able to describe the financial management of a company.
5	The Legal Framework	• Extract from a land leasing agreement • Transcript of a contractual discussion between lawyers	• Contract language • Terms and conditions	To be able to identify and describe the main elements of a contract.
6	Environmental Impacts	• Introduction to the Agenda 21 declaration • Statements from a company website showing their green credentials	• Environmental issues • Impact scenarios	To be able to describe how an oil and gas company manages environmental impacts.
7	The Development of an Oil and Gas Company	• The history of Standard Oil Company • Online encyclopedia account of the break-up of the company	• Company history, development and performance • Spin-offs, acquisitions and mergers	To be able to describe the history and development of a company.
8	Financing an Oil and Gas Operation	• Press release about a loan to a Hungarian oil and gas company	• Raising and lending finance • Targets	To be able to describe an investment made by a company and how it was financed.

Section 1
The Business of Oil and Gas

1 A Global Giant

Oil and gas is a global industry. As a worker in this industry you are part of a vast international network. As you travel with your work, you will meet people from many different countries and will need to be able to discuss issues affecting the oil and gas industry. This unit focuses on the vocabulary which will help you to converse in English about those issues.

Focused reading

1 Word pairs

Read this introduction to the oil and gas industry, which you might find in any online encyclopedia. Notice the pairs of words, e.g., 'oil and gas'. Make a note of any others you find.

Mozilla Firefox Start Page - Mozilla Firefox
File Edit View History Bookmarks Tools Help
http://www.oil&gasindustry.com
Most Visited Getting Started Latest Headlines

The oil & gas industry

The oil and gas industry is one of the largest in the world. It is made up of a small number of conglomerates and a very large number of smaller independent companies. Generally speaking, the industry is divided into two areas. Upstream operations involve exploration and production, often known as E&P, while downstream operations involve transportation, refining and sales of products to consumers.

Relationships between companies can be very complex, with large companies owning multiple subsidiaries around the world. In some cases, they compete for market share, in others they can be suppliers and customers of one another. They make joint ventures with other large companies for specific projects and there are frequent mergers and aquisitions.

Oil and gas production takes place onshore and offshore. Oil and gas companies are engaged in a constant search to discover new reserves and to obtain extraction rights for proved reserves. Construction of drilling wells and rigs is an enormous risk and companies need to calculate supply and demand in order to be certain that an operation will be profitable before it comes on-stream.

2 Getting the facts straight

Read these extracts from Internet websites relating to the oil and gas industry. Which of these statements are:

a purely factual?
b intended to show the importance of the oil and gas industry to the global economy?
c intended to promote one particular oil company?

1 'As the global economy evolves and expands, it presents enormous opportunities for the energy industry. We estimate that the world's oil and gas needs will grow 40% by 2030' (Rex W. Tillerson, Chairman and CEO, ExxonMobil).

2 'Proved reserves of oil are generally taken to be those quantities that geological and engineering information indicates with reasonable certainty can be recovered in the future from known reservoirs under existing economic and geological conditions' (www.bp.com).

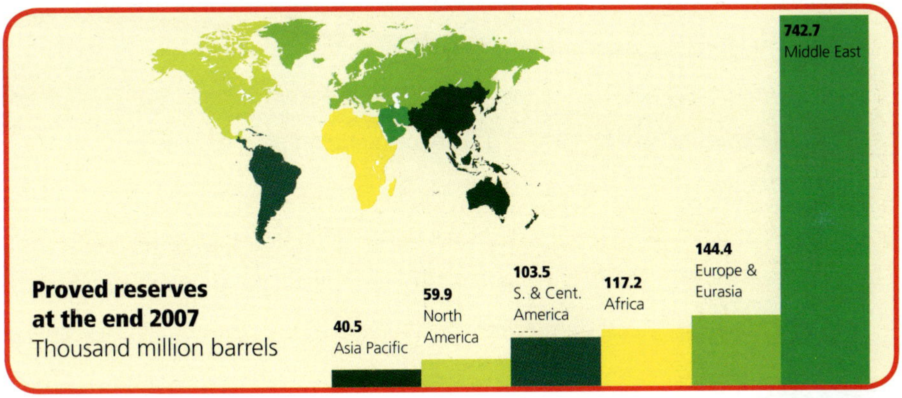

3 'Without oil and natural gas, quality of life would decline and people in developing nations would not be able to improve their standard of living' (www.spe.org).

4 'Strong demand from oil and gas producers has pushed rates for offshore drilling rigs to record highs, prompting many companies to build new rigs to meet the strong demand' (www.reuters.com).

5 'Shell E&P is engaged in the upstream activities of acquiring, exploring, developing and producing oil and gas' (www.shell.com).

3 Website accuracy

In many online encyclopedias, readers can edit information themselves if they believe the information is false. The following facts were correct in October 2008. Use a search engine, such as Google, to check and update any information which is incorrect or out of date.

1 The world produces and consumes 75 million barrels of oil a day.
2 The USA consumes 25% of the world's supply of oil.
3 China is the second largest consumer.
4 Sixty-eight per cent of the world's oil reserves are in the Middle East.
5 The price of a barrel of oil quadrupled from US$3 to $12 in the 1973 oil crisis.
6 The Thunder Horse platform in the Gulf of Mexico was hit by two hurricanes in 2005.
7 The biggest oilfield in the world is in Venezuela.
8 Oil and gas conglomerates employ 70% of their resources in upstream operations.

Vocabulary-building activities

1

Partnerships and prepositions

A *word partnership* is where two or more words are used together often enough to be considered as a fixed expression. Prepositions, or linking words, have an important role to play in word partnerships.

Here are a number of word partnerships in sentences. Write either of the prepositions 'of' or 'for' in each space, and highlight the part of each sentence you think is a word partnership. The first is done for you.

1 Demand for oil will continue to increase until 2015.
2 The market ________ liquefied natural gas is growing steadily.
3 Consumption ________ fossil fuels has declined slightly in recent years.
4 The supply ________ North Sea oil can be maintained until around 2024.
5 Transportation ________ oil presents many security risks.
6 Reserves ________ natural gas are being found almost daily.
7 Drilling ________ oil and gas is a major part of our upstream operations.
8 The extraction ________ oil at the Burgan field began in 1946.
9 Exploration ________ oil and gas has been going on for over a hundred years.
10 Production ________ gas from deep-water reserves is our main area of expertise.

2

Word families

A *word family* is where a number of words can be made from the same root word. Complete this table with words from these families, which are taken from Focused reading tasks 1–3, on the previous pages.

Verb	Noun (abstract)	Noun (person)
operate	operation	operator
extract		
	competition	
	acquisition	
		consumer
produce		
explore		
subcontract		
	supply	
transport		

3 Fill in the blanks

BP is an example of an oil and gas company operating globally. Below is an extract from its website describing its operations in Europe. There are a number of words missing from the document. Fill in the blanks with the right word from each group of words given in the table.

Mozilla Firefox Start Page - Mozilla Firefox

http://www.bp.com

bp

London is where BP's corporate headquarters are **(1)** ________________, and the UK is therefore a centre for many of its mainstream business functions. The UK is also home to three of BP's major global research and technology **(2)** ________________.

Our **(3)** ________________ and production business in Europe covers the North Sea – both the UK and Norway – and also The Netherlands. In Russia, we have an important joint venture through our 50% **(4)** ________________ of TNK-BP, a major oil company with the majority of its assets in Russia. We are involved in a number of E&P projects in Azerbaijan, and are leading the Baku-Tbilisi-Ceyhan (BTC) pipeline project.

Refining and marketing activities are spread throughout Europe, with BP owning or having a stake in nine **(5)** ________________ across the region. BP retail sites are a common sight in several European countries. In Germany, we **(6)** ________________ under the ARAL brand. We also sell lubricants and other oil products in Europe to both **(7)** ________________ and business customers.

locate	located	locating	location	locations
group	grouped	grouping	group	groups
explore	explored	exploring	explorer	exploration
own	owned	owning	owner	ownership
refine	refined	refining	refinery	refineries
market	marketed	marketing	market	
consume	consumed	consuming	consumer	consumers

Refresh

Using one of the strategies for consolidating and storing new vocabulary recommended in the Introduction, refresh what you have learnt in this unit.

Put it to work

Describe aspects of the oil and gas industry in your country of origin.

Section 1
The Business of Oil and Gas

2 Presenting the Company

Whether you are selling the services or products of your company, presenting to shareholders or orienting new employees, it is important to be confident and positive in speaking or writing about your company and its place in the oil and gas industry. This unit focuses on the vocabulary which will help you to present your company with confidence in English.

Focused reading

1 **A positive spin**

For a company's website to be effective, it needs to paint a positive picture. Read this extract from the website of LCM Oil & Gas Ltd. Note down some of the words and phrases used to describe the company and its operations in a positive way.

http://www.lcmoil.com

LCM Oil & Gas

... has wide recognition and a firm position in the sector of Oil & Gas Services in Libya. We provide tailor-made services in Construction & Maintenance (CMD) and Project Management Services (PMS) to all major operating oil and gas companies in the country. Since we started, we have grown consistently, and this success can be attributed to our commitment to working to the highest European standards with regards to communications, quality, safety, logistics and legal issues. LCM Oil & Gas Ltd has been operational in Libya since 1993. In 2003, LCM Oil & Gas Ltd was registered in the UK and the head office was implemented in London. Today, the head office, which was ISO 9001 accredited in 2004, supervises the entire operation of LCM Oil & Gas Libya. Over the past three years, our continuous growth and stabilization have helped us gain recognition in Europe. Our presence at worldwide events has spread our name and given us the opportunity to liaise with European partners and bring cutting-edge technology to the Libyan market.

2 Profiling right

Visit your company's or a competitor's website and study the company profile, its products and services, values and commitments. Which of the following are emphasized?

- Commitment to the environment
- Products or services
- Financial performance
- Marketing strategy
- Expertise in the field
- Management strengths
- History and track record
- Research and Development (R&D)
- Size
- International profile
- Employee welfare
- Other topics

3 Presenting to your audience

Read this slide from a PowerPoint presentation.

1 What is the aim of the presentation, do you think?
2 Who is the audience for the presentation?

ROC Performance Highlights

Upstream

- 6.4% production growth, well above industry average 2.4%
- Replaced 137% of production with new oil reserves
- Nine new licences acquired as part of strategy for long-term renewal

Downstream

- Refining throughput increased by 10%
- Two major refinery upgrades completed
- Focused expansion of retail network

Portfolio management

- Assets of three companies sold

Finance, corporate

- Strong revenues and profit performance
- Significant dividend recommended
- Corporate restructuring

Vocabulary-building activities

1 Going global

1 Complete this extract from a company website, using one of the words or phrases in the box.

market leader
joint ventures
rapid improvements
multinational
contract out

2 Which of the points in Focused reading 2 are emphasized in this extract?

Google - Windows Internet Explorer
http://www.multinationaloil.com

We are a **(1)** ______________ company, operating in 135 countries around the world. Our work is divided into upstream and downstream operations, with the majority of our turnover coming from E&P. We are the **(2)** ______________ in gas production, with a share of around 31%. We have seen **(3)** ______________ in technology in recent years and we now **(4)** ______________ a lot of our work to specialists and often make **(5)** ______________ with other companies in the same field.

2 Making the pitch

Which nine of the following 14 phrases would be most likely to appear in a presentation by the sales and marketing director of an oilfield services company?

a wide range of services
founded in 1979
commitment to excellence
key competitive advantages
environmentally sound
further to our discussions
kind regards
grouped into four geographical areas
leading-edge technology
proven track record
looking forward to meeting you
put you through
THE FIRST ITEM ON THE AGENDA
vital to our success

3 Winning the client

This is an introduction to a presentation by the marketing director of Cranford Oilfield Services to a potential client – a state-owned oil and gas company in the Middle East. Use some of the phrases on the previous page to complete the introduction.

Good afternoon, gentlemen. My name is Peter Ashling and I am the marketing director of Cranford Oilfield Services. The company was **(1)** ____________________ as a seismographic operation in the North Sea. Since that time, we have grown considerably and the company is now **(2)** ____________________ : The Americas, Europe and CIS, the Far East, Africa and the Middle East. We have a **(3)** ____________________ in most upstream operations and provide **(4)** ____________________ to our clients around the world. One of our **(5)** ____________________ is our flexibility. We are small enough to move quickly, but large enough to offer genuine cost-effectiveness. The company's policies are both customer-focused and **(6)** ____________________ . Our 3,000 staff have a dedication and **(7)** ____________________ which are second to none and **(8)** ____________________ in this very dynamic field.

4 Writing up your notes

Using the presentation slide in Focused reading 3, write a paragraph about the performance of ROC. Start like this:

In our upstream division, we had oil production growth of 6.4%, well above the industry average of 2.4%. We replaced ...

Refresh

Refresh the vocabulary you have learnt in this unit.

Put it to work

Prepare an outline PowerPoint presentation of your company for a new client, including an introduction, bullet-point company profile and positive conclusion. Try to use some of the positive language you have added to your vocabulary from this unit.

Glossary

Unit 1

Essential oil and gas vocabulary	
oilfield	where oil can be found and produced
upstream / downstream	the two areas of the oil and gas industry: upstream is exploration and production (E&P); downstream starts with distribution to refineries and finishes with sales of products to end-users
onshore / offshore	gas and oil can be found underground (onshore) or underwater (offshore)
proved (or proven) reserves	reserves where oil and gas can be produced with significant technical and economic certainty
come on-stream	when oil starts to flow from a field, the field is said to 'come on-stream'
to extract oil (extraction)	to get oil from the ground
explore (exploration)	to look for oil and gas
recover (recovery)	to find and produce oil or gas from a field
platform	where offshore exploration and drilling for oil and gas is done
producer (to produce / production)	the company that produces (drills for) oil or gas
operations (to operate / operator)	activities of work and business, such as exploration, extraction, transportation, etc.
transportation (to transport)	carrying goods (for example, oil or gas) from one place to another
Essential business vocabulary	
giant (multinational / super-major)	different terms for a very large company
competitor (to compete / competition)	another company operating in the same business as your company
supplier (supplies / to supply)	a company which sells products or services to your company
subcontractor (to subcontract)	a company that works for another company on contract
subsidiary	a company owned by another company or group
headquarters	the main offices of a company or group
a joint venture	a partnership between two companies to carry out a specific operation
supply and demand	in economics, the supply of a product is the amount that can be produced; the demand for a product is the amount that is needed by consumers
market share	the percentage (%) of the market a company has
consumer (to consume / consumption)	the final user of oil and gas products
mergers and acquisitions (to merge / acquire)	a merger is when two companies join together financially; an acquisition is when one company acquires (buys) another company
integration (to integrate)	when two systems or ways of working mix together
quadruple	to increase by a factor of four, as compared with double (× 2) and triple (× 3)

Unit 2

Company performance and strength	
strengths	the strong points (as opposed to the weaknesses)
key competitive advantages	very important advantages compared with other companies in the same business
commitment	a very positive attitude to something
expertise	a very high level of ability and knowledge
flexibility	the ability to change quickly and easily
cost-effectiveness	good value for money
dedication	giving completely your time, energy, etc.
customer-focused	doing everything possible to help the customer
Company development	
market leader	the biggest company in a sector of an industry
to contract out / a contractor	a lot of E&P operations are contracted out, i.e., given to smaller companies to do; the company which is contracted out is the contractor
turnover	the money a company makes from its sales
the company was founded	the company was started
has grown consistently	has developed at the same speed over a long period
continuous growth	increase over a long period
worldwide	international, global
spread our name	made our name known in many places
gain recognition	to become well known
accredited	given official approval
performance	what you or your company has done in terms of sales, profit, etc.
success	the positive things that you or your company has done
track record	the company's development in the market or your development in your career
renewal	change and improvement
expansion	another word for growth

Section 1
The Business of Oil and Gas

3 The Marketing of Oil and Gas Services

In order to sell its services, an oil and gas company needs to be clear about what it is selling (the product), how much its customers will pay (the price), where its services will be offered (place) and how it will present information about its services (promotion). This is the business of marketing. This unit focuses on the vocabulary which will help you to market your company's services effectively in English.

Focused reading

1

The 'four Ps'
In marketing, there are 'four Ps': Product, Price, Place and Promotion. Which of the four words go in each space?

> To create the right marketing mix, businesses have to meet the following conditions:
>
> - The goods must be in the right **(1)** ____________________ at the right time. Making sure that the goods arrive when and where they are wanted is an important operation.
> - The **(2)** ____________________ must be right. Consumers will need to buy in large numbers to produce a good profit.
> - The **(3)** ____________________ has to have the right features – for example, it must look good and work well.
> - The target group needs to be made aware of the existence and availability of the product through **(4)** ____________________.

Source: The Times 100 (www.thetimes100.co.uk)

2 Job opportunity

Read the following advertisement. Which of the four Ps is the main focus of the job?

> ### Marketing Communications Leader
>
> Maximize the company's image and brand awareness in the oil and gas exploration and production (upstream) market through exhibitions, advertising, press releases, e-newsletters and other key marketing communication activities. This position is focused on creating the marketing communication strategy and content of the company. The individual will be responsible for identifying needs and designing the content of creative sales/marketing materials, with a specific focus on advertising campaigns and media relations, press releases and articles, and high-level content of events. This person will work with the oil and gas marketing communications headquarters team for the final implementation of the communications projects.

3 Who's who?

In this telephone conversation, which of the four Ps is being discussed? Which of the two men works in the marketing department?

BRIAN: Hi Keith, how's it going?

KEITH: Fine, thanks – really busy, we're planning a lot of new campaigns.

BRIAN: That's good news. Listen, about the Khazakstan bid.

KEITH: Yes, how can I help?

BRIAN: Well, we need to set our prices and I was wondering if you could help out.

KEITH: Sure – who else is bidding, do you know?

BRIAN: Not for certain, but I have a feeling there'll be plenty of competition.

KEITH: So we'll need to be pretty sensitive. We may need to discount quite heavily.

BRIAN: Maybe. But I know that they would rather have good quality than something cheap and shoddy.

KEITH: What level of service are they looking for?

BRIAN: Well, I think platinum will probably be too pricey for them, but I think they'll go for gold, if we twist their arms.

KEITH: Yes, we don't want to price them out of the market, do we?

BRIAN: No, definitely not.

KEITH: OK, Brian – can you send me details and I'll give it some thought.

Vocabulary-building activities

1

Following the conversation

Read these statements and questions. Which of the four Ps is being discussed?

1 We need to update the Oil Tools website as soon as possible.

2 That's a very big discount. Do we need to offer that much?

3 The company offers a range of well-logging services for every environment.

4 Is there any way we can cut out the middleman and work directly with the client?

5 The new assembly on the control line can save several hours of rig time.

6 There's been a delay in printing the Deep-water Drilling brochure.

7 The bits are manufactured in the Far East and shipped throughout the world.

2

'Marketing speak'

Are these sentences true (T) or false (F) for your company?

		T	F
1	We attend exhibitions and trade fairs around the world.		
2	Our products or services are very price sensitive.		
3	We often offer a discount to our customers.		
4	We do a lot of advertising in specialist magazines.		
5	Our website is our main method of promotion.		
6	We issue press releases to international media organizations.		
7	Our brand awareness is very good.		
8	We change our products and services very frequently.		
9	We offer a wide range of products and services.		
10	We try to keep our prices as low as possible.		
11	There is a lot of competition in our field.		
12	We have price-fixing agreements with our competitors.		

3 The right price

Complete the sentences below using one of the following phrases.

price sensitive   set the price   pricey   price them out of the market
up our prices   price-fixing   agree a price   giving them a good price

1 We'll need to ______________ if we want to make a profit on the deal.

2 By ______________ we'll be able to insist on better payment terms.

3 I wouldn't ______________ too low – they'll think our quality is poor.

4 Oil and gas services are becoming very ______________ as competition increases.

5 I'm afraid these diamond bits are much too ______________.

6 I don't think we should go any higher. We don't want to ______________.

7 The big oil companies were accused of ______________.

8 If we fail to ______________, we'll have to find another supplier.

4 The product family

Complete these sentences with one of the following words.

produce   producers   producing   production   productive
productivity   products

1 We need to improve the ______________ of the drilling crew. They're just not working hard enough.

2 It was a very ______________ meeting. We made a lot of good decisions.

3 Demand is very high at the moment. Can we increase ______________ by around 10%?

4 A number of African countries have started to ______________ oil and gas recently.

5 Saudi Arabia is the biggest oil- ______________ country in the world.

6 Refined petroleum ______________ are usually grouped into three categories based on the way the crude oil is distilled.

7 OGP is the International Association of Oil and Gas ______________ (see www.ogp.org.uk).

Refresh

Refresh the vocabulary you have learnt in this unit.

Put it to work

Describe the products or services your company sells, and how these are priced and promoted.

4 Oil and Gas Accountancy

The nature of the oil and gas industry presents particular challenges for accountants. This unit focuses on vocabulary which will help you to talk or write about the financial management of your company in English.

Focused reading

1 **Basic accountancy terms**
Read these dictionary definitions. Which statement would show the value of a building or a large piece of equipment?

Balance Sheet A statement of the financial position of the company at a particular time, such as the end of the financial year or the end of a quarter, showing the company's assets and liabilities.

Profit and Loss Account (P&L) A statement showing the income and expenditure of a company over a period of time, usually one calendar year, balanced to show a final profit or loss.

2 Methods of accounting

This passage is about two different methods of accounting used by oil and gas companies – 'successful efforts' and 'full-cost accounting'. Read the passage. For each method, list the items which normally go in an oil and gas company's balance sheet, and list the items found in a P&L account.

There is some disagreement among experts over the correct way to account for oil and gas properties, but companies normally follow Generally Accepted Accounting Principles (either UK or US) and use a method known as 'successful efforts'. In this method, all property acquisitions, successful exploratory wells, all development costs (including development of dry holes) and support equipment and facilities are capitalized, and are therefore treated as assets in the company's balance sheet.

On the other hand, in this method, unsuccessful exploratory wells are charged to expense at the time it is agreed that the wells are non-productive, and recorded in the profit and loss account. In the alternative method – 'full-cost accounting' – all operations are capitalized whether successful or unsuccessful.

Items such as overheads and direct costs incurred in E&P are normally placed in the P&L account, but the treatment of items such as depreciation, depletion of any unproved reserves, or the costs of clean-up operations, varies widely from company to company.

In an industry where cross-border mergers and acquisitions are commonplace, factors such as national differences of accountancy principles, frequent changes to legislation, the use of estimation and currency fluctuations combine to make oil and gas accountancy a highly complex operation.

3 Feasibility study

This is a table from a slide presented in a meeting as part of a feasibility study into the construction of an oil and gas industrial park near the port of Aberdeen in Scotland. How much will be invested before any construction begins?

	£ million
Purchase of land and estate	450,000
Preparation and development of site	75,000
Civil works, structures and buildings	420,000
Production machinery and equipment	25,000
Environmental protection facility	70,000
Miscellaneous fixed assets	96,000
Contingencies (10%)	113,600
Total initial fixed investments	**1,249,600**
Pre-production expenditures and contingencies	16,500
Working capital requirements	50,000
Total initial investment costs	**1,316,100**

Vocabulary-building activities

1

Phrase recognition

Remove one word from each group and reorder the words to form a phrase.

Example:
account | and | loss | ~~balance~~ | profit   profit and loss account

1 incur | costs | facilities | unnecessary
2 exploratory | varies | unsuccessful | wells
3 to | charged | fluctuation | expense
4 reserves | of | overheads | depletion | unproved
5 and | mergers | acquisitions | capitalized | cross-border
6 sophisticated | operation | liabilities | highly

1 ______
2 ______
3 ______
4 ______
5 ______
6 ______

2

Accounts/Accounting/Accountancy

Complete these sentences using a word or phrase based on the root word 'account'.

1 We should have a word with our ______ about the new tax legislation.
2 A profit and loss ______ shows the difference between income and expenditure.
3 The ______ department has transferred to a new building.
4 My MBA included a whole semester studying ______.
5 The ______ show that we made a slight loss last year.
6 How do you ______ ______ depreciation in your company? (a two-word phrase)
7 ______ principles vary from one country to another.
8 The board of directors is ______ ______ the shareholders. (a two-word phrase)

3 Cost analysis

Look at the table in Focused reading 3. Are these interpretations of the costs presented there true (T) or false (F)?

		T	F
1	The total cost of buying, preparing and building on the land is around half of the total fixed investments.		
2	Not all of the costs can be forecast accurately and so an extra amount has been included.		
3	There is a very large amount of equipment needed on this site.		
4	An amount has been included to cover expenses after the site is operational.		
5	Approximately 45% of the initial investment costs are capital expenditure.		

4 Getting the numbers right

Write out in words the numbers from the table in Focused reading 3 in the blank column below, and then practise saying them aloud to yourself.

£ million	
450,000	
75,000	
420,000	
25,000	
70,000	
96,000	
113,600	
1,249,600	
16,500	
50,000	
1,316,100	

Refresh

Refresh the vocabulary you have learnt in this unit.

Put it to work

Describe a budget or financial plan for a project with which you have been involved. Describe how the plan worked in practice.

Glossary

Unit 3

Promotion	
customer / client	the person or company who buys your products or services
target group	the people you want to sell your products or services to
identify needs	understand what your customers want
brand awareness	how well people know your company and its products or services
company's image	what people think of your company and its products or services
advertising (advertise / advertisement)	promotion through magazines, television and other media (an advertisement is a single item of advertising)
issue a press release	to send information to a newspaper, magazine, etc.
advertising campaign	a series of promotion activities to advertise the company or a specific product or service
attend an exhibition or trade fair	go to an event where companies promote their products and services
media relations	contacts with the press, television, etc.
update the website	change the website with new information
Price	
to agree a price	find a price which is right for both sides
to set a price	establish a price for a product
to up the price (= to raise the price)	increase the price
price sensitive	demand changes a lot if the price only changes a little
to price a customer out of the market	the price becomes too high for the customer to consider
pricey	expensive
to discount the price	offer a special reduced price
price-fixing	illegal price agreement between suppliers
Other marketing language	
marketing mix	a marketing term to describe the way a company sells its products or services, focusing on a mix of product, place, price and promotion (the four Ps)
features	the important points which make a product different from other products
the availability of the product	where or when or in what quantity the product can be bought
implementation of a project	when a project is put into practice
level of service	related to price; companies sometimes offer different standards of service
cut out the middleman	work directly with the customer, without using an intermediary

Unit 4

Accountancy terms	
statement	a financial report made public by a company
to account for	to describe or record in financial terms
accountant (accountancy)	a person who keeps the financial records (the accounts) (accountancy = the profession of an accountant, what he or she does)
accounts	the financial records
balance sheet	a statement of the financial position of a company at a particular time
profit and loss account	a statement of the money a company makes (or loses) in a particular year
expenses / expenditure	the money a company spends
capitalize	to invest money in; to account as capital expenditure rather than day-to-day expenditure
overheads	the indirect cost of a company's business
depreciation	the reduction in value of an asset over the life of the asset
fluctuation	when the value moves up and down
contingencies	unexpected events
incur costs	to have to pay sums of money
charged to expenses	put into the expenses account
development costs	costs of developing a well
Assets	
asset	an item of property owned by the company; a positive figure on the balance sheet
purchase land (and estate)	buy the area where building work will be done
facilities	buildings with a specific function
to invest in (investment)	to spend money on something, hoping to get money back from it
machinery and equipment	the machines needed for a company or oil installation
fixed assets	property or machinery which is owned and not for resale
liabilities	the opposite of assets – negative items on the balance sheet
civil works	the kind of work done by a civil engineer, including, for example, roads, bridges, tunnels, wells, etc.
structures	a general word for buildings, machines, works, facilities, etc.

5 The Legal Framework

Multinational oil and gas companies encounter local, national and international law in every area of their upstream and downstream operations. Hundreds of contracts are drawn up and signed every day. This unit focuses on vocabulary which will help you to recognize and understand the main elements of an English language contract.

Focused reading

1 Standard terms

This is an extract from a typical contract. Who are the two parties to the contract?

Leasing Agreement

It is agreed that this lease shall remain in force for a term of FIVE (5) YEARS from the date of this lease and as long thereafter as oil or gas of whatsoever nature or kind is produced from said leased premises or on acreage pooled therewith, or drilling operations are continued as hereinafter provided.

If, at the expiration of the primary term of this lease, oil or gas is not being produced on the leased premises or on acreage pooled therewith, but Lessee is then engaged in drilling or reworking operations thereon, then this lease shall continue in force so long as operations are being continuously prosecuted on the leased premises or on acreage pooled therewith; and operations shall be considered to be continuously prosecuted if not more than One Hundred Twenty (120) days shall elapse between the completion of abandonment of one well and the beginning of operations for the drilling of a subsequent well.

If, after discovery of oil or gas on said land or on acreage pooled therewith, the production thereof should cease from any cause after the primary term, this lease shall not terminate if Lessee commences additional drilling or reworking operations within One Hundred Twenty (120) days from date of cessation of production or from date of completion of dry hole.

If oil or gas shall be discovered and produced as a result of such operations at or after the expiration of the primary term of this lease, this lease shall continue in force so long as oil or gas is produced from the leased premises or on acreage pooled therewith.

Source: Onecle, Inc. (www.onecle.com)

2 Following the conversation

In the following discussion, two lawyers are negotiating some of the terms of the contract on the previous page on behalf of the two parties involved.

1 Highlight the key points they discuss in the text of the contract on the previous page.

A: OK, so we're settled on five years as the term of the contract. What happens if we start producing just as the contract expires? Can we carry on?
B: Yes, we'll need to have a completion period.
A: And if we're no longer producing at all – what happens then?
B: Well, what would you suggest?
A: My proposal would be that we have another six months to start producing, and if we don't, then the lease expires.
B: Six months seems rather a lot. What about three?
A: OK. Let's call it 120 days. As long as we do any kind of work on the land within 120 days of the end of the five years, then the lease goes on. And if we need to finish off production, then we go on after that 120 days. How does that sound?
B: And then we negotiate a new lease. Is that what you have in mind?
A: I guess so. Let's draft this section and see how it looks.
B: Sure, good idea.

2 Note the expressions used for proposing and counter-proposing.
3 Practise reading the conversation.

Vocabulary-building activities

1

Précis

Read the contract shown in Focused reading 1 again. Now read these summaries of each of the four paragraphs. One of these summaries is inaccurate. Which one is it and why?

Paragraph 1	Under certain conditions, the producer can continue to operate after the end of the primary term.
Paragraph 2	If oil is not produced within 120 days of the five-year term, then the contract expires.
Paragraph 3	The producer can resume production at any time within 120 days of the expiry of the primary term.
Paragraph 4	The contract continues as long as oil is being produced.

2

Plain English

Look again at the contract in Focused reading 1. Find the following example phrase:

said leased premises or on acreage pooled <u>therewith</u> ...

This is an example of contract language. It means 'on the leased premises just mentioned or on acreage pooled with the leased premises'.

1 Can you find other examples of words beginning with 'there' or 'here' and explain what they mean?
2 Note down any other typical contract language that you find in the document.

3

A legal problem

Here is a situation where there was a problem with a contract. Put these stages of the contract into the most likely order.

1 Lawyers for each party checked the contract.
2 One party broke the terms of the contract.
3 One party paid compensation to the other.
4 One party sued the other for breach of contract.
5 The two parties adhered to the terms of the contract for some time.
6 The two parties drew up a contract.
7 The two parties negotiated the terms of an agreement.
8 The two parties signed the contract.

4 Contract language

Arrange these words from the contract into three vocabulary topic columns.

premises acreage lease expiration term in force
discovered produced pooled abandonment completion terminate
cessation of production land

Contract	The drilling area	E&P operations

5 Fill in the blanks

Complete the following sentences using some of the words or phrases from the previous exercise.

1 We had to ______________ the contracts of three crew members when they were caught smoking on the rig.

2 Shell and the Kazakh government have a production-sharing agreement on several ______________ in the south of the country.

3 After ______________, the well is abandoned.

4 The contract will remain ______________ until the ______________ of the well.

5 The ______________ of the contract can be extended with the agreement of both parties.

Refresh

Refresh the vocabulary you have learnt in this unit.

Put it to work

Refer to an English contract from your files or a sample contract from the Internet. Identify the parties to the contract, and summarize its main terms and conditions.

Section 1
The Business of Oil and Gas

6 Environmental Impacts

In recent decades, following the introduction of new laws and regulations, management of environmental impacts has become a higher priority for oil and gas companies. This unit focuses on the vocabulary which will help you to discuss, in English, environmental issues of concern to the oil and gas industry.

Focused reading

1

An environmental agreement

Environmental issues came to the fore for the industry in 1992 at the Earth Summit in Rio de Janeiro, and were formally recognized in a declaration known as Agenda 21. Read this passage about the Agenda 21 declaration. What kind of environmental impacts recognized by the declaration are represented in the three pictures?

Agenda 21

As long ago as 1992, at the Earth Summit in Rio de Janeiro, in a declaration known as Agenda 21, it was recognized that major industries such as oil and gas had an adverse impact on the environment. Issues such as sustainable land management, protection of the oceans, safer use of toxic chemicals and the management of waste, became crucial considerations for boards of directors around the world. In oil and gas exploration, there are several possible side effects, such as oil spills, damaged land, accidents and fires, and incidents of air and water pollution.

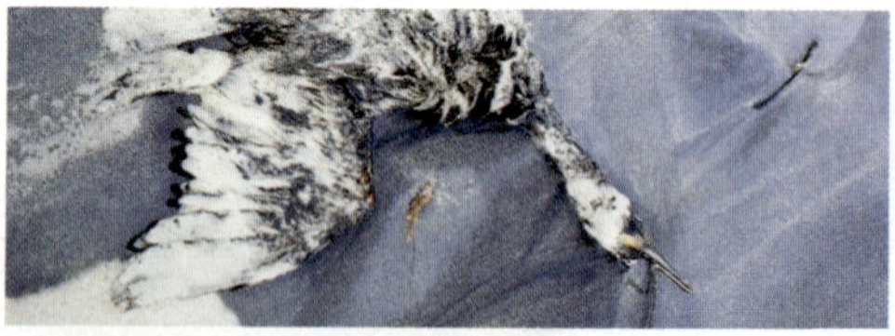

2 Adjective or adverb?

The following statements are extracts from the website of an international petroleum company, found in their Environmental Mission Statement.

1 Can you think of examples of words that might go in the spaces?
2 In each space, would you expect to find an adjective or an adverb?

Google - Windows Internet Explorer

http://www.gpoil.com

GP

Home Contact News

- GP aims to reduce the environmental impact of its operations by cutting waste, emissions and discharges, and by using energy ______________________.

- ______________________ air quality is one of the most topical environmental issues, ______________________ for large population centres. GP is helping to address this issue through the provision of ______________________ transport fuel in major cities worldwide. However, emissions from large operating sites are still a concern related to local or regional air quality.

- We operate in a variety of different environments around the world, from arctic to tropical, desert to deep sea. Many, if not most, of these areas are ______________________ and are often areas recognized to be of ______________________ biodiversity value.

- Recycling and reuse are now perceived as among the most ______________________ ways that the public, local authorities and industry can work together to help protect the environment.

- When projects reach the end of their lifecycles, we follow an approach designed to exercise ______________________ stewardship of the environment and leave a positive legacy in which sites can have a productive future use.

- GP is committed to delivering ______________________ commercial business improvement and greenhouse gas reductions through energy efficiency.

Vocabulary-building activities

1

Potential impacts

In the box below are words and phrases describing some of the potential environmental impacts of the oil and gas industry.

noise   soil contamination   greenhouse gases   water pollution   spillages
leaks   disturbance to wildlife   damage to habitat   global warming
visual intrusion   erosion

1 Which three do you think are the highest priority when *onshore* drilling?
2 Which three do you think are the highest priority when *offshore* drilling?

2

Impact scenarios

In the following situations, which of the potential environmental impacts in the box above need to be considered?

1 Oil has been discovered in an area of the Arctic where a rare species of whale lives.
2 In a storm, an oil tanker has sunk, losing its cargo of 30 million gallons of oil.
3 An onshore oil rig is under construction in beautiful countryside, close to a small, peaceful village.
4 Flaring during offshore production causes around 30% of the industry's CO_2 emissions.
5 Oil and gas companies remove land to create sites for wells, roads and other facilities. The work exposes bare soil to wind and rainfall.

3

Fill in the blanks

Complete the statements from the website extracts in Focused reading 2 using these words and phrases:

basic   cleaner   continuous   efficiently   environmentally sensitive
high   particularly   poor   responsible

4 Complete the sentence

Make complete sentences by choosing a word or phrase from each column. Draw in the links.

We operate in environmentally	spillages	to our environmental policies.
We must reduce carbon	improvement	in joules per annum.
It is important to avoid	sensitive	caused by flaring.
We measure energy	emissions	areas of the desert.
We aim for continuous	efficiency	by handling toxic materials with care.

5 Oil spill response

Complete this information about Norway's preparations for oil spills in its northern seas using these verb phrases.

are being used   has invested   can be pursued   have been drawn
has been made   have been expanded

- Norway's oil industry **(1)** ______________ NOK300 million in new clean-up equipment in recent years.
- With the latest oil-detecting radar and infra-red sensors, clean-ups **(2)** ______________ in darkness or with poor visibility.
- A substantial new commitment **(3)** ______________ to developing collection technology through an international collaboration.
- Landing crafts **(4)** ______________ to transport clean-up resources.
- Training and recruitment of personnel involved in oil company and public sector response organizations **(5)** ______________.
- Detailed maps of the coastal zone **(6)** ______________ to improve local response in the coastal and shore zones.

Refresh

Refresh the vocabulary you have learnt in this unit.

Put it to work

Either:
Make a list of what you, your team and your company do to protect the environment in your work. Make a separate list of things that you are not yet doing, but could consider.
Or:
Have you ever been involved in or witnessed an environmental incident? Describe what happened.

Glossary

Unit 5

Contract language	
lawyer	a person who has studied law and practises law as a profession
agree / disagree (agreement / disagreement)	to accept (or not) a proposal or contract
to draw up, or draft, a contract	to prepare a contract (a draft contract is not the final contract)
to sign a contract	to write your name on a contract to show that you agree with it
parties to the contract	the people who sign the contract
breach of contract	when one party breaks the contract
remain in force	to continue as a legal agreement
expire	to cease to be a legal agreement, usually because the legally applicable period is passed
negotiate a contract	to find an agreement between two or more parties
propose / counter-propose	to give an idea or to reject and give an alternative idea
on behalf of	a lawyer works for (or 'on behalf of') his clients
lease (lessee / lessor)	a contract to rent land or property for a period of time
term of the contract	the period of time a contract lasts
acreage	a piece of land (1 acre = approximately 4,000 square metres) for oil and gas operations
premises	buildings or land where a company operates
compensation	money paid by someone who breaks a contract to the other party in the contract
adhere to (e.g., the terms of the contract)	follow correctly (e.g., the terms of the contract)
sue	take legal action against someone who breaks a contract
E&P operations	
complete (completion)	after a well has been drilled, it is prepared for production (or completed)
abandon (abandonment)	to leave or quit; if a well has ceased producing or discovered to be dry, it is abandoned
discover (discovery)	to find or locate
to cease (cessation)	to stop completely
elapse	a period of time passes or elapses before something else happens
production sharing	when two or more companies produce together
crew members	people working together in a team
resume	start again

Unit 6

Environmental issues	
environmental impacts	things that can have an effect on the environment
adverse impact	negative effect
sustainable land management	managing land so that it can be used in the same way for a long time
waste	unwanted products from industry
crucial	very important
side effects	secondary effects of an industrial operation
oil spill (or spillage)	when oil is lost from a tanker or pipe
damaged land	land where there has been an adverse environmental impact
emissions (or discharges)	waste gases from an industrial operation
address the issue	find a solution to the problem
biodiversity	where there is a large number of different species of animal, plant, etc.
recycling	using materials again
protect the environment	make sure that the environment is not damaged
environmental stewardship	taking responsibility for protecting the environment
greenhouse gases	gases which cause the Earth to become hotter and contribute to global warming
energy efficiency	using energy without waste
soil contamination	damage to the material underground
leaks	when a pipe is damaged, oil leaks come from the pipe
disturbance to wildlife	when animals cannot live their normal lives because of pollution, etc.
visual intrusion	when something looks ugly, especially in a beautiful area
soil erosion	when soil is lost, either naturally or by careless industrial operations
flaring	a process where waste gas is burnt

Section 1
The Business of Oil and Gas

7 The Development of an Oil and Gas Company

When you are doing business, a very common topic of conversation is the development and growth of your company. Companies develop in many ways: they develop new products and services; they grow through mergers and acquisitions of other companies; they develop partnerships with other companies and with suppliers. This unit focuses on the vocabulary which will help you to converse in English about your company's development.

Focused reading

1

Standard Oil

Read the history of Standard Oil below. Which of these adjectives could you use to describe J.D. Rockefeller or Standard Oil?

efficient tiny weak badly managed ruthless powerful
innovative ordinary clever old-fashioned wealthy bankrupt
competitive lucky lazy

The History of Standard Oil

Standard Oil was a huge, integrated, oil-producing, transporting, refining and marketing company. Established in 1870, it operated as a major company trust and was one of the world's first and largest multinational corporations until it was dissolved by the United States Supreme Court in 1911. John D. Rockefeller was a founder, dominant partner and major shareholder, and the company made him a billionaire and eventually the world's richest man.

Standard Oil's market position was initially established through an emphasis on efficiency and responsibility. While most companies dumped gasoline in rivers, Standard used it to fuel its machines. While other companies' refineries piled up mountains of heavy waste, Rockefeller found ways to sell it.

In response to state laws trying to limit the scale of companies, Rockefeller and his partners developed new ways of organizing, to manage their fast-growing enterprise effectively. In 1882, they combined their disparate companies, spread across dozens of states, and created a single group of trustees. This organization proved so successful that other giant enterprises adopted this 'trust' form.

2

Mergers, acquisitions and spin-offs

Read the following brief account of the break-up of the Standard Oil group from an online encyclopedia and identify at least one merger, one acquisition and one spin-off operation.

www.online-encyclopedia.com

STANDARD OIL

The break-up of the Standard Oil group

- In 1911, the US Justice Department sued the Standard Oil group under the federal antitrust law and ordered its break-up into 34 companies.
- The successor companies from Standard Oil's break-up form the core (or Seven Sisters) of today's US oil industry. They include: Standard Oil of New Jersey (SONJ) or Esso (S.O. or Eastern States, Standard Oil) – renamed Exxon, now part of ExxonMobil.
- Standard Oil of New York – or Socono – merged with Vacuum – renamed Mobil, now part of ExxonMobil.
- Standard Oil of California – or Socal – renamed Chevron, became ChevronTexaco, but returned to Chevron.
- Standard Oil of Indiana – or Stanolind, renamed Amoco (American Oil Co.) – now part of BP.
- Standard's Atlantic and the independent company Richfield merged to form Atlantic Richfield or ARCO, now part of BP. Atlantic Richfield operations were spun off and bought by Sunoco.
- Standard Oil of Kentucky – or Kyso – was acquired by Standard Oil of California – currently Chevron.
- Continental Oil Company – or Conoco – now part of ConocoPhillips.
- Standard Oil of Ohio – or Sohio – now part of BP.

Vocabulary-building activities

1 Key oil and gas dates

Match the following events with the dates given below. If you don't know the answer, take a guess or research further at www.geohelp.net/world.html.

1859	1870	1879	1895	1903	1931	1938	1954	1973	1989	1997

1. Anglo-Persian Oil Company was renamed British Petroleum.
2. Colonel Edwin Drake struck oil 69 feet below the surface of the ground in Titusville, Pennsylvania.
3. Conrad and Marcel Schlumberger successfully identified the presence of oil in a formation by measuring resistivity.
4. *Exxon Valdez* ran aground in Prince William Sound, Alaska.
5. Ford Motor Company was founded.
6. J.D. Rockefeller formed Standard Oil (Ohio), controlling 10% of American oil refining.
7. Oil prices quadrupled from $2.90 a barrel to $11.65 a barrel, following an Arab oil embargo on oil exports to the US.
8. Oil was discovered in Kuwait and Saudi Arabia.
9. The combustion engine was invented.
10. The Kyoto agreement was made with the aim of limiting greenhouse gases.
11. Thomas Edison invented the electric light bulb.

2 Slide show

Read the following extracts from a slide show presentation about the recent development of an oil and gas company and match them to the graphs below. Two of the statements are not illustrated.

1 Our share price has risen steadily for the last ten years.
2 The number of accidents has fallen sharply since we introduced continuous shifts in 2005.
3 There was a slight drop in output in the third quarter of last year.
4 Profits remained stable despite the fall in oil prices.
5 Growth in the industry is expected to decline in the next twenty years.
6 We increased our market share by around 4% in 2008.
7 Development activity has gone up slightly since the recent rises in oil prices.
8 Our sales reached a peak in November.

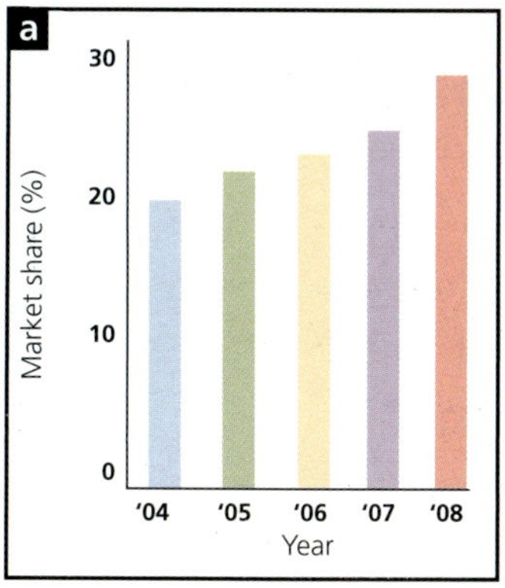

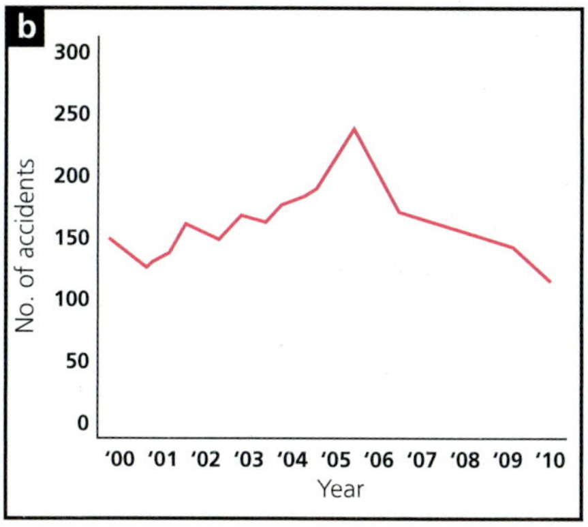

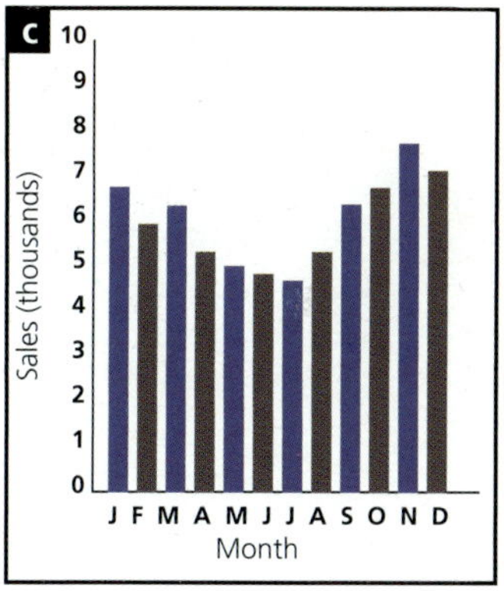

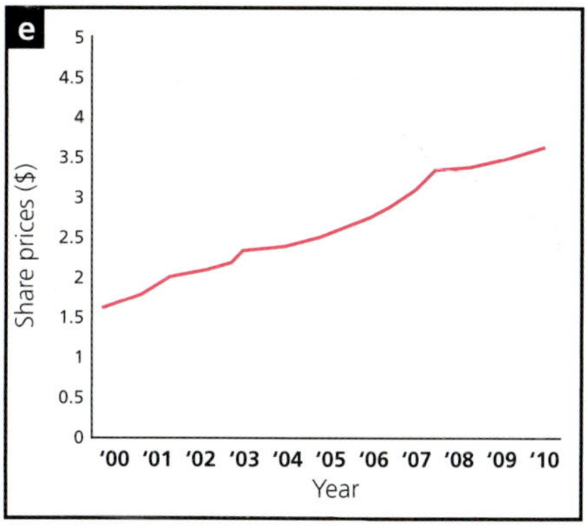

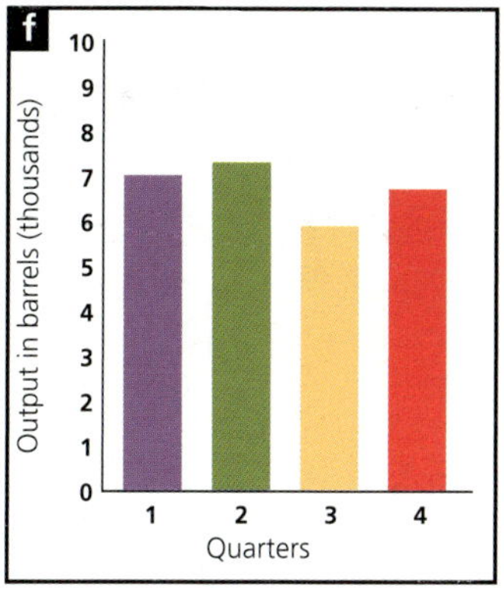

Refresh

Refresh the vocabulary you have learnt in this unit.

Put it to work

Describe some important events in your company's history and how it has developed since it was incorporated.

Section 1

The Business of Oil and Gas

8 Financing an Oil and Gas Operation

Oil and gas operations are very expensive. Companies need to enter into complex agreements with banks and other financial institutions to ensure that they are able to finance their operations and achieve their aims. This unit focuses on the vocabulary which will help you to discuss company finance in English.

Focused reading

1

A conditional loan

MOL, the Hungarian Oil and Gas Company, needed to raise a lot of money. Read the press release, opposite, from the European Bank for Reconstruction and Development (EBRD).

1 Explain the significance of the following figures:
   a 3 years
   b €150 million
   c €750 million
   d €600 million
   e €226 million

2 Explain the relationship between Michel-Marc Delcommune and Alain Pilloux.

Press release, 28 July

€150 million loan to MOL to cut emissions and improve environment

European Bank
for Reconstruction and Development

Hungarian Oil and Gas Company to reach EU norms within three years

The EBRD is lending MOL Hungarian Oil and Gas Company €150 million for investments aimed at upgrading the environmental performance of the company's operations in Hungary. It is part of a €750 million financing package, alongside a €600 million loan MOL is raising from a syndicate of commercial banks for general corporate purposes.

Hungary has accepted all EU standards and conditions relevant to the oil and gas sector. As a result, MOL aims to reach compliance with the applicable norms within three years. To achieve this goal, the company plans to invest a total of €226 million – including the €150 million from the EBRD – over this period.

Under the loan terms, MOL is required to adhere to the Environmental Action Plan agreed with the EBRD and ensure that all relevant milestones are met. For its future acquisitions, there is a requirement to carry out adequate due diligence, and develop and implement action plans to bring new facilities and subsidiaries into compliance with international and local environmental standards and practices. This should have significant environmental, health and safety benefits.

'MOL welcomes the completion of its landmark €750 million financing package and, as a key part of this, the loan agreement with the EBRD. These transactions demonstrate the support of the banks for MOL's strategy to become the leading oil company in central Europe and to maintain quality leadership, while meeting high environmental standards,' said Michel-Marc Delcommune, Chief Financial Officer of MOL.

Alain Pilloux, EBRD Business Group Director for Central Europe, said the Bank is pleased to provide financing intended to assist MOL in substantially improving its environmental performance.

© European Bank for Reconstruction and Development. Website: www.ebrd.com

Vocabulary-building activities

1

Misreporting the facts

Look again at the EBRD press release. The information in six of the following sentences has been misreported and needs to be corrected. Use a red pen to make the necessary amendments.

1. The loan of €150 million is aimed at maintaining environmental standards.
2. Under the terms of the loan from EBRD, MOL aims to reach compliance with the applicable norms.
3. MOL expects to reach compliance within five years.
4. The EU requires MOL to adhere to the Environmental Action Plan.
5. MOL has not agreed to carry out adequate due diligence.
6. The banks would like to discourage MOL from adhering to environmental norms.
7. MOL intends to reduce the quality of its leadership while achieving low safety standards.

2

Responsibilities in financing

Which is more likely to do the following: a bank or an oil and gas company? Write the phrases in one of the three columns below.

lend money borrow money obtain a financing package provide financing
sign a loan agreement raise money offer funding give a loan secure a loan
get financing negotiate a loan set interest rates

Bank	Oil and gas company	Both parties

3 Targets

In English, there are a number of verbs that are used in the context of discussions about setting and achieving targets. In the following sentences, there are a number of right answers. You can fill in the blanks using two, three or, in one of the sentences, four of the possible words and phrases given below. For each sentence, which words might be used?

achieve   adhere to   comply with   ensure   fulfil
maintain   meet   reach   satisfy

1 If you ______________ your sales targets, you will get a good bonus.
2 Hopefully, we can ______________ the objectives we discussed at our last meeting.
3 We need to ______________ a good level of profitability in the UAE.
4 He was sacked because he did not ______________ the safety regulations.
5 The goals I would like us to ______________ are as follows …
6 The company has a legal duty to ______________ the environmental standards of the oil and gas industry.
7 Unfortunately, this equipment does not ______________ all our criteria.

4 Conditions

There are many different ways to express conditions in English. Which of the words in the box can replace the underlined phrase in the sentence below?

after   as long as   assuming   before   during   if
provided that   providing   unless   when   while

The EBRD is going to lend MOL €150 million <u>on condition that</u> they agree to improve their environmental performance.

Refresh

Refresh the vocabulary you have learnt in this unit.

Put it to work

Either:
Describe an investment made by your company and how it was financed, or how you would go about raising funds for a new development initiative.
Or:
What targets have you been set that you have to achieve in your job? How do these relate to your company's goals for the future?

Glossary

Unit 7

Adjectives for describing a company	
efficient	working in an organized and effective way (as opposed to in an inefficient way)
innovative	always trying to use the newest technology and ideas
powerful	very strong in the market
huge	very big (as opposed to tiny, or very small)
competitive	always trying to get more business and take market share
dominant	the leading company in the market
Verbs for describing the history of a company	
found (or establish)	start the company
discover	find for the first time
invent	make for the first time
rename	give a different name
spin-off	separate from the main company
break-up	divide into smaller companies
merge	join together into a bigger company
acquire	buy another company
spread	grow geographically
form (or create)	to make a new company
Describing trends in a company's performance	
rise	go up (as opposed to fall, or 'go down')
increase	go up (as opposed to decrease, or 'go down')
slightly	a little
steadily	a similar amount over a long period
sharply	a lot very quickly
a drop	a fall
decline	go down
remain stable	stay the same
reach a peak	arrive at the highest point

Unit 8

Raising and lending finance	
to lend	to give something that you expect to get back (the opposite of 'borrow')
interest rate	the percentage you pay when you borrow money
funding	providing money for a business project
financing package	financial support given by a bank in different forms
a loan agreement	an agreement to borrow money from a bank
to raise money	borrow money to finance a business project
Targets and duties	
targets	what you want to have in the future (usually clear and precise); you *attain* or *hit* a target
goals	what you want to have in the future (a little less precise than targets); you *realize* or *reach* (or score in the context of sports) a goal
objectives	what you want to have in the future (another word for a goal); you *realize* or *reach* an objective as well
aims	what you want to have in the future (a little less precise than goals and objectives); you *achieve* an aim
achieve	to reach the level you want
aimed at (… ing)	with the aim of doing something
ensure	to make sure
reach compliance with	to arrive at a level where you are following the regulations
applicable norms	the standards which are accepted
adhere to the rules	to follow the rules
comply with (e.g., the regulations, the law, etc.)	to follow correctly (e.g., the regulations, the law, etc.)
meet relevant milestones	to achieve the interim goals you have agreed
due diligence	a process of checking the finances of a company which is the target of a takeover
carry out (e.g., due diligence, or other tasks)	perform (e.g., due diligence, or other tasks)
fulfil a task	to perform a task
bring the company into compliance	to ensure that the company follows the rules
discourage	to make someone less willing to do something (as opposed to 'encourage')
satisfy (e.g., the criteria, the standards, etc.)	to do what is expected by (e.g., the criteria, the standards, etc.)
requirement	something you have to do by a contract or law

Section 2 Contents

Working on a Rig

	Titles	Focused Reading	Vocabulary Areas	Aim(s)
1	Rig Locations	• Descriptions of four different oilfield locations	• Oil rig locations • Oilfield terrains and weather conditions	To be able to describe the location of an oil and gas field.
2	Offshore Platforms and Vessels	• Descriptions of different types of platform construction • Extract about rig support vessels • Diagram of an FPSO	• Rig and platform construction	To be able to compare different rigs or platforms.
3	Jobs on the Rig	• Recruitment agency descriptions of the work of drillers and toolpushers • Job advertisement for field engineers from a trade magazine	• Job roles, responsibilities, tasks and skills	To be able to describe your job and those of your colleagues.
4	Working Conditions	• Magazine article extracts on working conditions offshore • Research project reports about shift work	• Working conditions • Verbs that indicate cause and effect relationships	To be able to describe daily routines on a rig.
5	Health and Safety	• Extracts from accident reports • Safety guidance document	• Risks on a rig • Safety equipment and clothing • Describing accidents	To be able to discuss health and safety risks and describe measures to avoid them.
6	Maintenance and Repair	• Job interview transcripts with candidates describing approaches to maintenance	• Maintenance, servicing and repair	To be able to describe the repair and maintenance of equipment on a rig.
7	Rig Erection and Dismantling	• Transcript of an interview between a journalist and a platform manager • Newspaper article about the decommissioning of a rig	• Rig erection • Rig dismantling	To be able to describe the erection and dismantling of a rig.
8	Rig Automation	• Extracts from a conference presentation • PowerPoint presentation slide about seabed rigs • Online encyclopedia entry about robotics in oil exploration and production	• Rig automation • New technologies • Making forecasts • Robotics	To be able to describe a robotic or unmanned system used on a rig.

1 Rig Locations

Oil and gas operations take place in some of the most remote and environmentally hostile regions in the world. Oil and gas rigs can be found at sea, in the desert, in remote corners of the jungle and on the polar ice caps. This unit focuses on vocabulary which will help you to describe in English the locations of different oil and gas fields.

Focused reading

1 Extreme challenges

Read the descriptions of rig locations around the world, opposite, and think about these questions as you read.

1. Which aspects of these rig locations do you find exciting and why?
2. Which aspects of their locations present challenges and in what respect?

Oil drilling on the Siberian tundra

Total has interests in a number of concessions in Murzuq basin, located in the south-west of Libya. The desert landscapes are extremely varied and exceptionally beautiful. Rainfall is rare (significant every 3–7 years) and occurs between October and March. The rest of the year is dry and hot (up to 50°C), with wind sandstorms in March and April. In addition, Murzuq is renowned for its numerous archaeological sites, from Palaeolithic to Neolithic times.

bp

Indonesia's Papua Province (formerly called Irian Jaya) on the western half of the island of New Guinea is the site of BP's Tangguh liquefied natural gas (LNG) facility. Indonesia is a tropical archipelago of around 17,500 islands. It has the richest diversity of flora and fauna of any nation on earth – an estimated 17% of the world's species – including an extraordinary number found only in one location. The country encompasses a vast variety of habitats, including 42 major terrestrial ecosystems and five coastal/marine ecosystems.

Oil is by far the most important mineral product in Alaska. Exploration for oil and gas is in progress in several land areas and on the outer continental shelf from the Gulf of Alaska to the Beaufort Sea coast. The 800-mile-long Trans Alaska Pipeline System (TAPS) is one of the largest pipeline systems in the world. It stretches from Prudhoe Bay on Alaska's North Slope, through rugged and beautiful terrain, to Valdez, the northernmost ice-free port in North America. Since pipeline start-up in 1977, Alyeska Pipeline Service Company, the operator of TAPS, has successfully transported over 15 billion barrels of oil.

Hurricane Dean reaches Belize

The first tropical storm of the year, Hurricane Dean, hit the Yucatan Peninsula this morning, with lashing rain and driving winds of up to 168 miles (270 kilometres) an hour. Petroleos Mexicanos, the state-owned oil company, evacuated almost 19,000 workers and ordered 437 offshore wells to be shut down. The hurricane is moving west through the Mexican jungle and is forecast to regain strength Tuesday night as it crosses the Bay of Campeche. Coastal areas are bracing themselves for extensive flooding and storm damage.

Vocabulary-building activities

1 The world's biggest

1 Complete the following sentences about the location of some of the world's biggest oilfields. Refer to a world atlas if you need to.

located between located at off the coast of along the north-east coast of in northern in the Norwegian sector of from the northern edge of

a Samotlor Field is the largest of Russia's oilfields, __________ Lake Samotlor in Urals Federal District, Siberia.

b Azeri-Chirag-Guneshli (ACG) is a large complex of oilfields in the Caspian Sea, about 120 kilometres __________ Azerbaijan.

c Ekofisk is an oilfield __________ the North Sea.

d Shaybah is about 40 km __________ the Rub' Al-Khali/Empty Quarter desert and is a major crude oil producing site in Saudi Arabia.

e Bolivar Coastal Field is the largest oilfield in South America, stretching 35 miles __________ Lake Maracaibo, Venezuela.

f The Athabasca Oil Sands are a large deposit of oil-rich bitumen, or extremely heavy crude oil, __________ Alberta, Canada.

g Daqing complex is the largest oilfield in the People's Republic of China (PRC), __________ the Songhua river and the Nunjiang river in Heilongjiang province.

2 Circle the prepositions used in each sentence.

The Athabasca Oil Sands

Daqing complex

Shaybah

Samotlor Field

2 All-weather operations

In which of the environments given in the column headings in the table below would you expect to meet the following weather conditions? Complete the table.

dry and arid   monsoons   driving winds   blizzards   freezing cold   sandstorms
gale-force winds   hurricanes   hot and humid   tropical cyclone   permafrost
lashing rain   ice-floes

Offshore	Desert	Jungle	Arctic
			permafrost

3 Reports from the frontiers

The following quotes describe rig locations and are reported by people who have worked on remote oilfields. What kind of environment was each of them working in, do you think? Use the four environments from the table above.

1. We were surrounded by 30-metre trees in all directions.
2. I have never known such cold weather, never above –25 degrees Celsius.
3. The sun was like a furnace. You had to be really careful of sunburn.
4. Fortunately, I don't suffer from seasickness, but many of the crew were ill that day.
5. Sometimes you woke up in the morning and there were drifts of snow all over the rig.
6. Working in a dust storm is really unpleasant. You can hardly breathe.
7. The typhoon struck at dawn. The waves were up to 25 metres high.
8. The worse thing was the insects. You had to cover yourself in cream.

Refresh

Using one of the strategies for consolidating and storing new vocabulary recommended in the Introduction, refresh what you have learnt in this unit.

Put it to work

Describe an oil and gas rig or an oilfield you have worked on, or one which you have visited – its location, the environment and the weather conditions.

2 Offshore Platforms and Vessels

A high proportion of oil and gas exploration and production is carried out offshore. Platform construction has developed over time to become better adapted to deal with the physical challenges of different offshore locations. This unit focuses on vocabulary which will help you to describe in English the different offshore platforms and vessels on which you or employees of your company work.

Focused reading

1 The right platform

Study the table describing the different platforms and vessels used for drilling offshore. Which of these are:

1 easy to move from one place to another?
2 probably fixed in one place for a long period of time?
3 designed for use in shallow water?
4 designed for use in deeper water?

Type of platform/vessel	Description
Drilling barge	A flat-bottomed rig used in shallow waters.
Jack-up rig	A rig which has two hulls and at least three vertical legs through the hulls.
Semi-submersible rig	A rectangular-shaped platform floated with pontoons below the surface of the water.
Drillship	A ship with a drilling rig mounted in the centre.
Gravity-base fixed production platform	A platform with a steel-reinforced concrete base, held to the seabed by gravity.
Steel-jacket platform	A platform which has steel legs that sit on the seabed, held by concrete piles.
Tension-leg platform	A platform held in position by heavy weights connected to hollow steel tubes.
Compliant platform	A light platform designed to sway with the wind, waves and currents.

2 Construction types

Some of these types of construction are shown in the pictures below. Read the descriptions again and see if you can match them up.

1

2
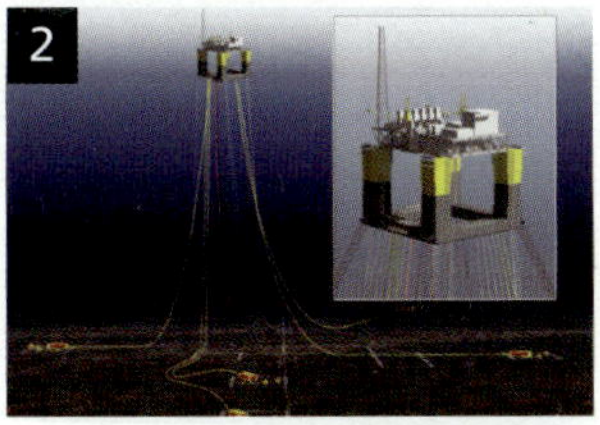

3

4

5

Global Lake Drilling (GLAD800) platform, Lake Titicaca, Bolivia (photo courtesy of DOSECC)

3 Support vessels

Frequently, oil companies use other support vessels attached to existing rigs. An example of this is a Floating Production, Storage and Offloading vessel (FPSO). Study the diagram below and read the following description of FPSOs. Why do oil companies use FPSO vessels?

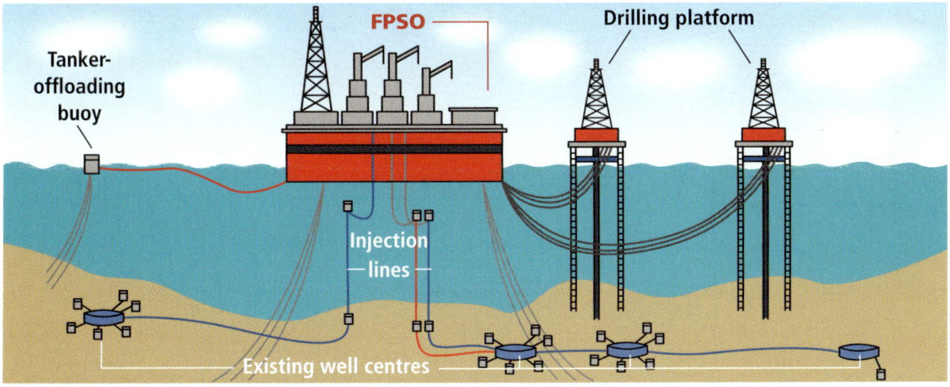

FPSO vessel

> The Floating Production, Storage and Offloading vessel is a type of floating tank system designed to take all of the oil or gas produced from a nearby platform or platforms, process it and store it until the oil or gas can be offloaded onto waiting tankers, or sent through a pipeline. Such vessels are particularly effective in remote or deep-water locations where seabed pipelines are not cost-effective. FPSOs eliminate the need to lay expensive long-distance pipelines from the oil well to an onshore terminal. They can also be used economically in smaller oilfields which can be exhausted in a few years and do not justify the expense of installing a fixed oil platform. Once the field is depleted, the FPSO can be moved to a new location.

Vocabulary-building activities

1 The features of a rig

Using words from the box below, label the features visible in the photograph. (Not all of the features given in the box are shown.)

jacket	piles	accommodation modules	derrick	drilling rig	flare stack	pontoons
service cranes	drilling pipe deck	heli-deck	store areas	ballast tanks	barge	hull

2 Form and function

Match the features in the box above with the functions listed below.

1 This is where the helicopter, bringing crew and supplies to the platform, lands and takes off.
2 These are used for lifting heavy equipment onto the rig.
3 Supplies are kept here until they are needed.
4 Waste emissions are burnt off through these.
5 These are filled with water to submerse the platform.
6 This structure supports the crown block and strings of a drilling rig.
7 This is the legs and structure that support the platform.
8 These are used to secure the platform to the seabed.
9 This is used to pull the platform into position.
10 This is where the crew sleep and relax when they are off-duty.
11 This is the equipment used for drilling.
12 This is where the sections of the drillstring are stored.

3 Know your rig

Read the text in Focused reading 2 and study the diagram of a FPSO again. Are the following statements true (T) or false (F)? Check the boxes as appropriate.

	T	F
1 An FPSO vessel is a small drilling rig.		
2 The FPSO can offload oil or gas to either a tanker or a pipeline.		
3 It can store oil produced on several different platforms.		
4 It is useful in shallow waters just off the coast.		
5 It is used as a replacement for seabed pipelines.		
6 It saves a lot of money in certain situations.		
7 It works well in fields where oil or gas is expected to last a long time.		
8 It is more cost-effective than a fixed platform in some situations.		
9 It is very mobile and flexible.		

4 Attraction of opposites

1 Find a word or phrase in this unit that means the opposite of the word in bold in the following sentences:

a This is where the helicopter **lands**.
b This type of rig is suitable for **deep** water.
c The crew are **on-duty** 12 hours per day.
d This is a **heavy** type of platform.
e The offloading vessel is a **fixed** system.
f First, the ballast tanks are **emptied**.
g It is a **horizontal** construction.
h It is a **solid** cylinder made of steel.
i We want the equipment to **sink**.

2 Make a negative word from these adjectives using 'in' or 'un'.

economical effective expected expensive flexible necessary suitable

Refresh

Refresh the vocabulary you have learnt in this unit.

Put it to work

Compare and contrast two different rigs or platforms you have visited. What are the differences and similarities between the two? How is the construction of each platform adapted to its location?

Glossary

Unit 1

Oil rig locations	
located	describes where a place is; used with different prepositions (e.g., in, on, at, between, etc.)
northern / southern / eastern / western	describes location with relation to an area or region of a country or city, for example
in the north / north-west / south / south-west, etc., of …	describes the location with relation to a place in a country or city, for example
on the edge	on the outer limit of a place
sector	an area of sea or land owned by a country or company
stretching	spreading over an area of land
Oilfield environments and weather conditions	
hurricane	a violent storm with very strong winds, often in the Atlantic Ocean
typhoon	a violent tropical storm with very strong winds, often in the Pacific or Indian Ocean
driving winds	very strong winds
lashing rain	very heavy rain
desert landscape	what you can see in the desert
sandstorms / dust storms	storms in the desert when the wind blows sand or dust everywhere
tropical archipelago	a group of islands and the sea around them
rugged terrain	rough, often mountainous landscape
arid	very dry
monsoon	very heavy rain falling at a particular time of year in the tropics
blizzard	very heavy snowfall
gale-force winds	very strong winds
cyclone	very strong winds moving in a circle
permafrost	land that is permanently frozen
ice-floe	a large area of ice floating in the sea
remote	far away from land or from places where people live
drifts of snow	occur when there is wind and heavy snow together
waves	the up-and-down movement across the sea

Unit 2

Rig and platform construction	
flat-bottomed	without a hull
shallow	not deep
hull	the shaped bottom part of a boat
rectangular-shaped	a rectangle has four sides; two sides are longer than the other two
mounted	fixed onto a flat surface
steel-reinforced	made stronger with steel
hollow	with nothing in the middle
sway	to move slowly from side to side
currents	the natural movements under the surface of the sea
float	stay on the surface of the water (as opposed to sink)
offload	to take goods from a ship or other vehicle
barge	a low boat used for carrying goods and equipment
piles	columns of concrete
crane	a machine with a long movable arm for lifting heavy objects
jacket	the legs and support for the platform
derrick	the structure that supports the equipment in a drilling rig
accommodation module	a pre-fabricated building where the crew of a platform live while they are working on the rig
flare stack	a vertical tower used to burn gases
land / take off	a plane lands when it comes down to earth and takes off when it leaves
heli-deck (or heli-pad)	a flat area for helicopters to land and take off
store areas	places where supplies and equipment are kept
offloading area	the area of the platform or support vessel where oil is transferred to tankers or pipelines
ballast tanks	the tanks on the legs of the platform; ballast helps something to remain stable on the water
buoy	a floating object used for marking areas of sea or for tying up vessels

Section 2
Working on a Rig

3 Jobs on the Rig

There are many different jobs on an oil and gas rig. Some are managerial, requiring sensitivity and high-level leadership skills. Some are technical and detail-oriented. Many are dirty and tiring manual jobs. This unit focuses on vocabulary which will help you to describe in English the jobs that you and your colleagues do on the oil rig.

Focused reading

1 Job descriptions

If you are looking for a job in the oil and gas industry, a good place to start is a specialist recruitment website. Read the following descriptions of two typical jobs on an oil and gas rig, which you might find on any oil industry recruitment website. Underline the key tasks of each job.

http://www.jobsinoilandgas.com

Drillers apply

Drillers have a hands-on role on the rig. Their tasks include moving, setting up and operating drilling equipment to drill holes for exploration. They may also move rigs onto site, monitor equipment, including how fast the drill is going, and assist in crane operations. Once drilling activity begins, a driller operates the brake and motor of the drill, and determines how fast and deep the drill needs to go. Throughout the day, drillers need to record drilling progress. Drillers also play a 'troubleshooting' role, determining and fixing any drilling-related problems that occur during their shift.

View all jobs

http://www.jobsinoilandgas.com

Toolpushers

apply

Toolpushers monitor the drilling operations or activity on board an oil and gas drilling rig. They relay their instructions to drillers on the rig via the telephone. They also manage personnel and equipment, and must have an understanding of all drilling processes. This knowledge allows them to set the drilling schedule, which defines the rate at which work will be undertaken and how much work will be completed in a shift. A toolpusher constantly monitors the schedule, making relevant changes when required. Toolpushers are also in charge of delegating work to the most appropriate employees. It is a highly responsible position and a huge variety of duties need to be undertaken on an oil rig, so careful consideration needs to be given to this allocation of work.

View all jobs

2

The right skills

Read this job advertisement for field engineers for an oil and gas service company working in the Middle East. Identify and write a summary of the most important competencies and areas of expertise needed for the job.

http://www.oilgasjobs.com

Field Engineers

Region: Middle East

Salary: Competitive

Join **OGTec** for a career in a booming, high-tech oilfield service company. OGTec Inc is seeking field engineers to take part in developing the company, by delivering the best services to our customers in the area of the Arabian Gulf States. You can expect personal development in a demanding work environment. It is challenging, exciting and rewarding. Expect every day to be different. You will experience a great trust in your capabilities and will be the first point of contact to our customers. You will also be responsible for very critical and expensive operations.

The overall task is to operate OGTec technology for oilfield service applications:

- Operate electronic and hydraulic equipment downhole.
- Consult with customer and suppliers to plan layout of equipment. Study blueprints, schematics, manuals and other specifications to determine installation and procedures.
- Install or oversee installation of equipment according to manufacturers' specifications.
- Operate systems to demonstrate equipment, utilizing knowledge of electronics and using standard test instruments and hand tools.
- Instruct and direct workers in servicing and repairing the equipment.
- Consult with engineering personnel to resolve unusual problems in system operation and maintenance.

apply

Vocabulary-building activities

1

Who's who on the rig

The following job titles and descriptions below describe just some of the many people who work on an oil rig.

driller installation manager roustabout assistant driller toolpusher company man crane operator chief engineer mud logger wireline operator roughneck

1 Match the job title to its description. Write your answers in the first column.
2 In the last column, write down whether it is a *manual* job, a *technical* job or a *management* job.

Job title	Job description	Job type
	represents the oil and gas operator	
	sets up and handles the rotary drilling equipment	
	is in charge of the platform	
	analyzes samples of rock cuttings	
	carries out drilling operations under supervision	
	handles supplies and equipment	
	operates the various electronic tools used in the well	
	sets up and operates lifting equipment	
	supervises operations on the rig	
	is responsible for overall operation of the drilling rig	
	handles pipes and other equipment for the rig	

2 Relate, communicate and adapt

Members of oil and gas crews must have strong relational and communication skills and must be highly adaptable. In which of the three skill categories – relational, communication or adaptability – would you place the following competencies?

1 Able to work with a wide variety of people
2 Able to work in a tightly-knit team
3 Able to work in harsh conditions
4 Able to meet the demands of the client
5 Able to absorb precise detailed information
6 Able to speak English clearly and effectively
7 Able to give clear, concise instructions
8 Able to motivate colleagues
9 Able to work long hours
10 Able to stay calm under pressure

3 Management responsibilities

These are some of the management tasks carried out by engineers, toolpushers, driller supervisors and other managers.

1 Which five are most important in your job?

handling paperwork   authorizing supply orders   monitoring the budget
checking expenditure   scheduling work   delegating work
requesting technical information   coordinating plans   managing day-to-day activities
overseeing the operation   ensuring excellent results   motivating other people

2 What other responsibilities do you have as a manager?

Put it to work

Either:
Imagine that you are moving to a new position and have to describe your job to a recruitment specialist in order for him to write a job advertisement. Describe your job, your responsibilities and the skills required.
Or:
If you weren't doing the job that you do, which other job on the rig would you most like to do and why? Describe the job, its responsibilities and the skills required.

Refresh

Refresh the vocabulary you have learnt in this unit.

Section 2
Working on a Rig

4 Working Conditions

Working conditions on an oil and gas rig can be arduous. When people are working long shifts in confined spaces, good teamwork and respect for one another are vital. The experience can be very rewarding. This unit focuses on vocabulary which will help you to describe daily working conditions on a rig.

Focused reading

1

Working conditions offshore

The magazine *Oil and Gas World* recently wrote an article on working conditions offshore. They interviewed two mud loggers; Hamish Thomson and Abi Reardon. Read the following extracts from the article. Which aspects of their working conditions are described in each extract?

Extract 1

Hamish Thomson

Hamish is contracted to work in the North Sea and so far has worked in Norway, Denmark and Holland. He has used this as an opportunity to travel throughout northern Europe when onshore. His employer provides services on oil rigs all over the world, so he may well have further opportunities for travel in the future. As mud loggers are in demand wherever there is oil, travel is viewed as one of the perks of the job.

Extract 2

'Mud loggers typically work two weeks offshore followed by two weeks at home. When I'm onshore, I am not actually 'at work', but I can be called offshore at any time, even at weekends. The only time I can't be called back offshore is when I am on booked annual leave. It's a bit like being on-call as, even though I might be at home, I still have to be prepared to go back offshore if required.'

Extract 3

Abi Reardon

Abi is one of a growing number of women working in the oil industry: currently between 10% and 15% of the working population on oil rigs are women. Oil and rig companies have gradually added separate facilities onto rigs to make it easier for women to join the industry. Abi says that she feels completely at ease working in an environment which is still predominantly male. 'I am treated first and foremost as someone who does a job, rather than as a woman,' she says.

Extract 4

'The work is always 12 hours on, 12 hours off when on a 'hitch' (shift) offshore or onshore. Each hitch usually lasts between two and four weeks, with an equal amount of time off between hitches. The shift start times also vary; either 6 a.m. to 6 p.m. for day shifts, or 6 p.m. to 6 a.m. for night shifts. Depending on where you work, living conditions can vary somewhat, and getting used to the working offshore lifestyle is difficult for some.'

2 Shift work

Here are the conclusions of two recent research projects focused on shift work.

Typically, a worker on an offshore installation works tours of between 14 and 21 days, and does shifts of 12 hours a day or night.

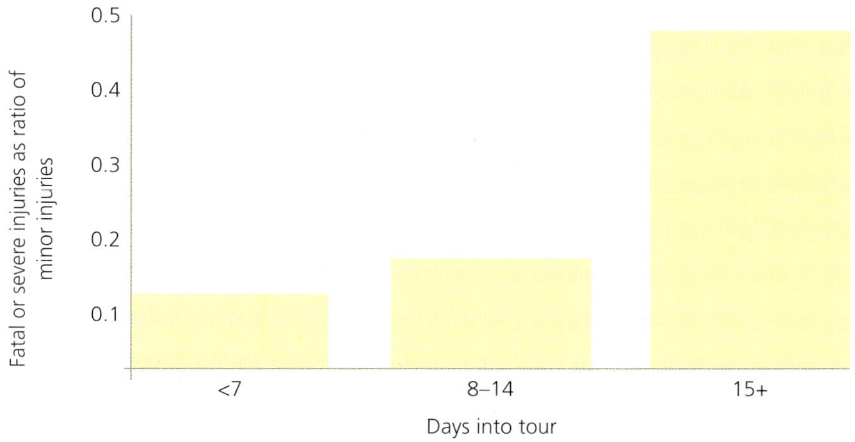

Research has shown that in terms of alertness, mood, sleep patterns and accidents, 14D shifts are best, but after a few days of adjustment, 14N shifts are almost as good. On the other hand, rollover shifts cause far greater health and safety risks.

14N = 14 continuous nights   14D = 14 continuous days
7N + 7D = 7 nights followed by 7 days (a rollover shift)
7D + 7N = 7 days followed by 7 nights (a rollover shift)

Based on this research, what recommendations would you give to the managers on a rig to ensure they safeguard employee health and safety?

Vocabulary-building activities

1

'Pros and cons'

Which of the following would you say are the positive aspects of working on a rig and which are the negative?

1 You can make a good living.
2 There is a strong team spirit.
3 The work can be hazardous.
4 Shift work is arduous.
5 Weather conditions are extreme.
6 You learn a variety of skills.
7 You get very dirty.
8 It is a predominantly male environment.
9 When you are onshore, you can be called back to the rig at any time.
10 You have to be away from family and friends for long periods.

2

Upsides and downsides

Complete these sentences so that they express your opinion about working conditions on a rig. You may wish to use phrases from Focused reading 1 to help you. (If you don't work on a rig, you can modify the sentences using 'would', as shown in the example below.)

Example:
The best thing about working on a rig would be that … you would get first-hand experience of nature's extremes.

The worst thing …
I (would) really like the fact that … and that …
What I (would) dislike most is …
Personally, I don't (wouldn't) mind (being / getting / having) …
I found (would find) it difficult to get used to (being / getting / having) …
But I (would) really enjoy …

3 Cause and effect

The following sentences describe an all-too-common sequence of causes and effects when it comes to shift work on a rig. Construct the sentences using forms of the verbs that follow. The first one is done for you.

(… can / could …)

cause	lead to	bring about	result in	expose rig workers to

1 Working long tours ▶ ___can lead to___ a lack of sleep.

2 A lack of sleep ▶ ______________ a loss of concentration.

3 A loss of concentration ▶ ______________ a higher frequency of mistakes.

4 A higher frequency of mistakes ▶ ______________ an increased risk of accidents.

5 An increased risk of accidents ▶ ______________ injury or even death.

4 Awake and alert

The conclusions of the research in Focused reading 2 show that there is a problem of alertness for workers on rollover shifts. This conclusion can be stated in different ways using the verbs or verb phrases below.

make	impair	reduce	cause	decline	affect	suffer	result in	deteriorate

1 Which of the verbs or verb phrases below can be used in the following sentences?

a Rollover shifts seem to ______________ your alertness.

b Rollover shifts seem to ______________ you less alert.

c Your alertness seems to ______________ during rollover shifts.

2 Two of the words in the box above cannot be used in these sentences. Which are they?

3 Can you make another sentence stating the same conclusion as above, which can include these other two words?

Refresh

Refresh the vocabulary you have learnt in this unit.

Put it to work

Describe daily working conditions for a worker on a rig. The hardships and challenges may be easier to identify than the benefits. Imagine you are a careers adviser 'selling' working for an oil and gas company in this kind of setting: why would a newcomer to the industry want to work on a rig?

Glossary

Unit 3

Describing skills and responsibilities	
a manual job	a job working with the hands (as opposed to a clerical or management job)
highly responsible	very responsible
the job requires … sensitivity, patience, etc.	you must have these skills in order to do this job
a wide variety of people	many different types of people
have a hands-on role	work very close to day-to-day operations
play a troubleshooting role	the work will involve solving day-to-day problems
fixing problems that occur	solving problems that happen
in charge of …	responsible for …
undertake a duty	agree to perform, or carry out, a task
allocation of work	deciding who does each job
competencies	skills and abilities
areas of expertise	parts of the job where you know a lot and are very experienced
a demanding environment	a difficult working environment
work in a tightly-knit team	work in a team in which the people are very close to each other
work in harsh conditions	work in very difficult conditions
absorb detailed information	take in and understand detailed information
Specific oil rig tasks and duties	
set up equipment	put the equipment into position and get it working (also 'install equipment')
monitor equipment	watch it to make sure that it works correctly
assist in crane operations	help with or take part in crane operations
determine how deep …	decide how deep …
record drilling progress	keep a note, or record, of drilling as it is done
relay instructions	pass instructions to other people
set the drilling schedule	decide what drilling will take place and when

Unit 4

Working conditions	
do shift work	to work in a job that is carried out in shifts, typically for the oil and gas industry, two or three shifts a day
12 hours on, 12 hours off	when you work for 12 hours and then have 12 hours when you don't work
hitch	another word for shift or tour
varies somewhat	is different to a degree; could vary by a little or a lot; not well-defined
perks of the job	the additional benefits of the job
on leave	on holiday, away from work
facilities	the things that you need to live on a rig (e.g., rooms, toilets, etc.)
predominantly male	mostly men work on the rig
positive / negative aspects	things about the job which are positive / negative
make a good living	get a good salary
hazardous work	dangerous work
arduous work	very hard work
a variety of skills	many different skills
I don't mind …	it's not a problem for me
to get used to … ing	it gets easier to accept
a lack of …	not enough
a loss of …	less than there was
alertness	a state where you can focus well and not make mistakes
mood	how you feel; when you are happy, you are in a good mood

Verbs describing cause and effect	
causes leads to brings about results in	words that can all be used in the structure: a → b (where 'b' is the result of 'a') e.g., tiredness causes / leads to / brings about / results in a loss of concentration
reduces impairs affects	words that can be used in the structure: a → b (where 'a' has a negative effect on 'b') e.g., tiredness reduces / impairs / affects concentration
deteriorates declines suffers	words that can be used in the structure: b ________ as a result of a e.g., concentration deteriorates / declines / suffers as a result of tiredness

Section 2
Working on a Rig

5 Health and Safety

Safety is a major concern on an oil and gas rig, as the environment poses many hazards. The risk of falling or of getting caught or hit by machinery is always present, so those on board need to be fully trained and alert to the dangers. This unit focuses on vocabulary which will help you to identify health and safety risks in English, and to describe measures that should be taken to avoid accidents.

Focused reading

1 Incident reports

Read the four extracts, opposite, from reports of incidents during oil and gas activities. Match each extract to one of the following summary descriptions.

1 An operator was hit by a pipe which had been forced from its assembly by gas pressure.
2 An operator was killed and another injured during an escape of poisonous gas.
3 Some operators were injured during a small fire caused by sudden ignition of a gas line.
4 Some operators were killed in a serious fire.

a

The plant housed separation equipment used to produce oil and natural gas from nearby wells. Workers were purging a two-mile pipeline between an oil and gas well and the facility when a vessel ruptured due to over-pressurization, releasing flammable material, which then ignited. Natural gas from the ruptured vessel produced a large fireball, which damaged nearby piping and released and ignited additional flammable materials. Four workers were killed in the accident.

b

At a well stimulation site, an operator needed to release excess carbon dioxide (CO_2) from a mobile tanker unit, known as a queen. The queen was parked between two other queens, so the excess gas could not be released directly out to the side. To redirect the venting gas to the front of the queen, the CO_2 operator constructed a pipe assembly using available piping components. When the CO_2 operator opened the discharge valve to release the excess CO_2, the pressure of the venting gas unthreaded the pipe assembly and it spun counter-clockwise. The pipe assembly struck the CO_2 operator, causing fatal injuries.

c

Three employees of JT Oilfield Construction Company were performing maintenance work on an oil separator unit when the accident occurred. One was on a ladder at the top of the tank while the other two were disconnecting a water line at the base of the unit. A gas line next to the water line became dislodged. The local Sheriff, David Masters, said the gas immediately ignited: 'The pipe was pointing down with about 40 pounds of pressure when it ignited.' Local Fire Chief, Grant Hogan, reported that the gas fire was not major. No oil ignited. 'It was more or less a flash fire from the raw gas and was put out quickly.'

d

A service rig company had been hired by an oil company to 'complete' a well. The service company believed the well to be 'sweet'. Prior to the accident, the producing formation had been acidized and two members of the crew had been requested to stay overnight and flow the well. At approximately 01:00, after all the acid flush water had been returned to surface, oil began to flow. At 02:30, after gauging the rig tank, one worker was overcome with H_2S. The second worker, who was sleeping at the time of the first knockdown, awoke at 03:30 and found the first worker down. While attempting to rescue the first worker, the second worker was knocked down. Both workers were found on the ground the next morning, one had expired, the other was unconscious, but alive, and has since recovered.

Vocabulary-building activities

1

Protective equipment

Identify and label the following pictures of safety equipment and clothing with the help of the health and safety guidance given below.

- Safety goggles should be worn to protect the eyes from loose cuttings.
- Respirator masks should be worn to avoid breathing in dust or fumes.
- Protective footwear should be worn during heavy lifting.
- Rubber gloves should be worn when working with chemicals.
- Earmuffs don't eliminate noise completely.
- A face shield is required when welding to prevent sparks from going in the eyes.
- High-visibility jackets and hard hats are compulsory on the rig.

1 ______________________

2 ______________________

3 ______________________

4 ______________________

5 ______________________

6 ______________________

7 ______________________

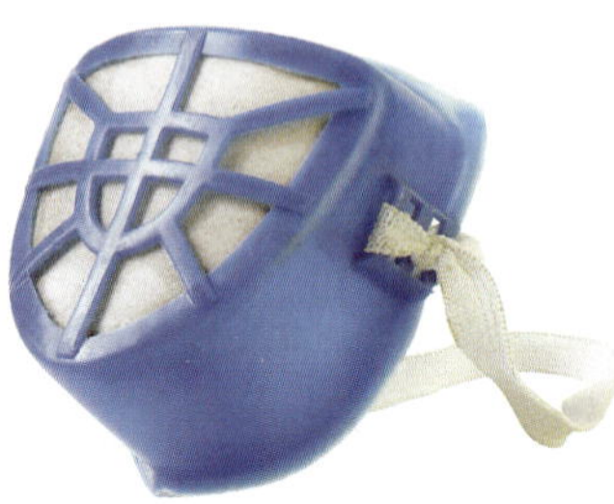

8 ______________________

2 Health and safety training

Read the safety advice about hydrogen sulphide training. Complete the advice using the phrases in the box.

detection methods | first-aid procedures | personal protective equipment | sources and hazards | symptoms | warnings

Comprehensive training should be provided for workers in H_2S operations. Example topics include:

- Identification of the characteristics, **(1)** ________________ of hydrogen sulphide.
- Proper use of the hydrogen sulphide **(2)** ________________ used on the site.
- Recognition of and proper response to hydrogen sulphide **(3)** ________________ at the workplace.
- **(4)** ________________ of hydrogen sulphide exposure.
- Proper rescue techniques and **(5)** ________________ to be used in a hydrogen sulphide exposure.
- Proper use and maintenance of **(6)** ________________. Demonstrated proficiency in using PPE.

3 Describing the accident

Use an insertion mark [^] to indicate where the words in brackets should be inserted. Write them above the insertion mark in the line space provided.

1 An operator was knocked unconscious. (young, pipeline)

2 Another operator was injured. (badly, experienced)

3 The drilling schedule was disrupted. (completely, first)

4 A storage area was destroyed. (chemicals, totally)

5 Some piping was damaged. (essential, slightly)

6 The fire crew was informed. (immediately, local)

7 The fire was extinguished. (fortunately, quickly)

Refresh

Refresh the vocabulary you have learnt in this unit.

Put it to work

Either:
Describe the health and safety measures used at a site where you have worked or one that you have visited.
Or:
Describe an accident that occurred at a site you know. What training have you had in order to deal with such an accident? Do you feel prepared?

6 Maintenance and Repair

The failure of a single item of machinery on an oil and gas rig can lead to time-consuming and expensive shutdowns, not to mention risks to operators. It is essential, therefore, that a programme of maintenance and repair is in place so that plant runs smoothly at all times. This unit focuses on vocabulary which will help you to identify and describe maintenance issues and the implementation of repairs in English.

Focused reading

1

A safe pair of hands

RigRepair is a service company specializing in the maintenance and repair of oil and gas rigs. Its Human Resources Director, Alvin Gibbs, is interviewing five candidates for an International Project Manager position. Each candidate has had experience of a different approach to maintenance.

Read the descriptions of their approaches to maintenance, opposite. What are the advantages and disadvantages of each? From the information, which candidate would you prefer for the job?

Anthony Johnson

I suppose the type of maintenance we do would be called 'corrective maintenance'. In other words, we're pretty casual about it and just take action when something breaks down. So far this has worked out OK. We're quite a small outfit and everybody works very closely with the equipment. You usually get a feeling when something is going wrong, and a quick check will show what needs mending or replacing. We certainly don't do anything systematic.

Faisal Riyad

We have a systematic procedure of preventive maintenance. This involves planned inspections of each item at specific times, and every two years we do a complete overhaul during a scheduled shutdown of the rig. We perform tests, measurements and adjustments; we replace or replenish lubricants and other fluids; and we replace worn parts so that faults can be corrected before they develop into major defects.

Pramana Batak

We have recently introduced an approach which is called 'predictive maintenance'. We found with preventive maintenance that we did a lot of unnecessary work and wasted time and money. We used spare parts before it was necessary, we threw away oil which had months of use left in it, we shut down the rig far too often and lost production. So, we've invested in special software for analyzing criticality and we use infrared devices to forecast potential failures.

Colin Freyer

I am a rope access technician. I work in a harness up to 70 metres above sea level. A lot of the maintenance work we do involves using ultrasonic equipment, which measures the thickness of metal. If metal is supporting a heavy load or transporting gas or liquid, after a while it will wear and tear, and it's very important that it's repaired if it becomes too thin. I guess it's quite dangerous work, but I get a big thrill from it.

Igor Borovsky

It is important for us to keep our older platforms producing and in good working condition, so we constantly need to monitor and modify our equipment. In the past, we carried out surveys and assessed the work required by taking hand-read measurements and making two-dimensional drawings. Nowadays, we can do this with a hand-held scanner to record and visualize complex components within construction projects and working installations. Instead of climbing ladders and around dangerous pipework with ropes and harnesses, a laser scanner allows the data to be collected remotely from a distance of up to 50 metres.

Vocabulary-building activities

1

Plugging the gaps

For each sentence, choose three of the five words to complete the sentence. Two of the words are not correct.

1 During shutdown, we ________________ the equipment for defects.
check, control, inspect, monitor, supervise

2 We ________________ many different tests and analyze the results using special software.
carry out, do, make, perform, take

3 Regular maintenance enables us to detect ________________ in the equipment.
flaws, damage, defects, errors, mistakes

4 We ________________ a lot of data to ensure the accuracy of the tests.
assess, collect, gather, measure, obtain

5 If we find a defect, we can ________________ the part.
remove, replenish, repair, replace, restore

6 Taking preventive ________________ saves a lot of money on expensive repairs.
action, activity, measures, maintenance, steps

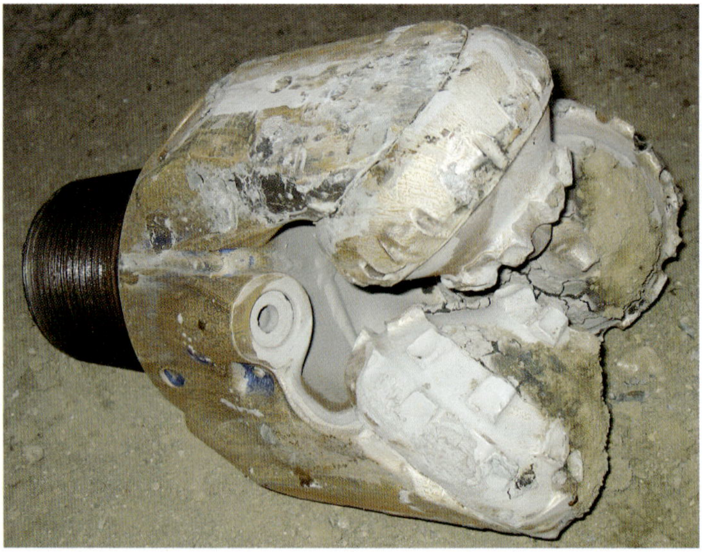

A worn drill bit

2 Maintaining and servicing

Match the correct definition to each of these types of maintenance.

breakdown maintenance   condition-based maintenance   corrective maintenance
predictive maintenance   preventive maintenance   reliability-centred maintenance
risk-based maintenance

1 Planned maintenance or servicing at regular intervals to avoid failures occurring in the future.

____

2 Maintenance carried out at the optimum time, based on forecasts of a machine's lifespan and failures that could happen during that time.

____

3 Maintenance carried out at the optimum time, based on analysis of events which have caused damage to this or other such machines in the past.

____

4 Maintenance carried out at the optimum time, based on an analysis of the current health of a machine and what could happen to it in the future.

____

5 Maintenance to determine, systematically and scientifically, what must be done to ensure that physical assets continue to function normally and as the users want them to.

____

6 Maintaining only when inspection shows that it is required.

____

7 Not maintaining at all, unless the machine has actually failed.

____

Refresh

Refresh the vocabulary you have learnt in this unit.

Put it to work

Either:
Describe the way that equipment is maintained at the site where you work or one that you have visited.
Or:
Describe the maintenance of a specific piece of machinery.

Glossary

Unit 5

Risks on an oil and gas rig	
poses many hazards	causes hazards or risks
alert to dangers	awake to the possible dangers
awareness (e.g., of dangers, risks, etc.)	knowing about the dangers
detection (of gas, etc.)	finding (gas, etc.) by smell or other means
emergency procedures	actions to be taken if an incident occurs
exposure to H_2S	when H_2S is breathed in or has contact with the skin
identification of a hazard	knowing that something will be dangerous
proficiency in using equipment	having the ability and knowledge to use the equipment
recognition (of dangers, etc.)	seeing and understanding (that something is dangerous, etc.)
symptoms of H_2S exposure	the physical results of exposure
warnings	spoken or written statements about something dangerous that could happen
protect from danger	make something safe
prevent from happening	make sure that something dangerous doesn't happen
Safety equipment and clothing	
face shield	worn over the whole face, with a Perspex mask
goggles	worn over the eyes
rubber gloves	worn on the hands
high-visibility jacket	worn to enable others to see you
protective footwear	shoes or boots worn to protect the feet
earmuffs	worn in a very noisy work environment
hard hat	worn to protect the top of the head
respirator mask	worn over the mouth to enable breathing in dusty conditions
Results of accidents	
the accident occurred	the accident happened
damaged	does not work or look like it did before the incident
ruptured	badly damaged where the equipment is split open
ignited	caught fire
destroyed	damaged so badly that it no longer exists
dislodged	moved from its original place
disrupted	when a procedure is interrupted so that it no longer works successfully
(fatally) injured	a person dies when they are fatally injured in an accident

Unit 6

Maintenance, servicing and repair	
failure / breakdown	when a machine has a problem and stops working
an item of machinery	a piece of machinery, or of a machine
shutdown (= outage)	when a plant, such as an oil rig, stops working for a time to allow maintenance and repair to take place
time-consuming	taking too much time
runs smoothly	operates without problems
mend	an informal word for repair
take corrective action / steps / measures	do something to repair a machine when it breaks down
take preventive action	do something to avoid a failure
overhaul	to completely maintain or repair a machine
carry out or perform a test	do a test
replace a part	to install a new part instead of a damaged or worn one
replenish (fluids, stocks, etc.)	to add more fluids or stocks when the level is low, so that the original quantity is achieved
worn parts	parts which have been used for a long time and need replacing
detect flaws / defects / damage	look for and find things which are wrong with a part or machine
predictive maintenance	forecasting or anticipating what will happen to a machine and maintaining at the best time possible
waste time	to use your time badly
spare parts	additional machine parts to replace existing ones
criticality	concerning the importance of a maintenance action: when you 'assess criticality', you decide how important the job is
rope access technician	an operator who works in parts of a rig with access only by rope
harness	a safety device for carrying operators when working at heights
thickness	a measurement of materials such as metal (thick is the opposite of thin)
wear and tear	damage caused by use of a part or machine
assess (the work required)	estimate or calculate (how much work is needed)
hand-read machine / hand-held machine	a machine read or held directly by the operator (i.e., not by a computer)
remotely	from a distance
collect / gather / obtain data	to get data by measuring, reading, etc.

Section 2
Working on a Rig

7 Rig Erection and Dismantling

Erecting an offshore oil and gas platform is a highly sophisticated, lengthy and expensive operation. Although the rig will be in use for many years, the dismantling and disposal of the rig once the field is depleted needs to be considered from the outset. This unit focuses on vocabulary which will help you to describe the erection and dismantling of a rig in English.

Focused reading

1 Platform construction

Alan Noughty is a journalist working for a local newspaper in Aberdeen. He is doing research for an article about developments in oil and gas platform construction since North Sea oil was discovered in the mid-60s. Read the transcript, below, of part of an interview he conducts with a platform manager, Jo Mcleod.

Which photo, opposite, should Alan include with his article and why?

AN: At what stage is the decision taken to build a platform?

JM: Well, platform fabrication is pretty much the first step after establishing that a field will be commercially viable.

AN: How long does it take?

JM: The jacket – that's the legs and support for the rig – can take at least two years to build. But having said that, fields in the North Sea are getting smaller these days, so we can often speed things up by using smaller specialist equipment attached to existing rigs. We can get those into operation in a few months.

AN: I see. And how do you get the jacket out here once it's built?

JM: It has to be towed out to sea by barge during a 'fine weather window' in summer and up-ended into position by flooding the ballast tanks in its legs and secured to the seabed with concrete piles. The deck support and production units are put in place by crane barges which can lift over 10,000 tonnes of equipment at once.

AN: That's incredible.

JM: Sometimes with concrete jackets the initial erection is carried out in a sheltered bay and the whole structure is towed out and secured. Nowadays, we tend to use steel which is much lighter.

AN: And how long does it take before the platform is up and running?

JM: Anything up to a year really. We have to hook up and test all the equipment ready for drilling and production. There can be a thousand people working on the platform before it comes on-stream!

AN: Amazing.

1

2

2 **Decommissioning**

This is another article written by Alan Noughty for his newspaper, *The Aberdeen Echo*. List the key phases of the decommissioning process and/or describe the work involved.

Local firm awarded contract to dismantle rigs

A local company has been awarded the country's biggest ever oil rig dismantling contract. Abbott UK will recycle and dispose of the North-West Lennon platform at its Blackthorn yard, in Hartford. Last night, Abbott Chief Executive, Peter Stephens, said: 'We are very pleased. We have been involved in the technical process of getting the contract for a couple of years. The North-West Lennon is a very big platform and dismantling it will mean two years of work.' The rig's topsides and steel jacket will be removed by heavy lift vessel and towed 130 km to Abbott's Hartford yard for disposal. BP awarded the offshore removal and onshore recycling of the rig to prime contractor Heering Marine Contractors, of Leiden, in the Netherlands. Work will be carried out by a range of specialist subsea, engineering and waste handling companies, including Abbott. The contract represents the largest remaining element of the North-West Lennon decommissioning project. Work already completed includes the well abandonment and topsides clean-up phases carried out after the field ceased production at the end of 2008. The pipeline decommissioning, debris clearance and site inspection is all that remains. BP is proud of its excellent safety performance record to date. 'Ensuring that we complete the offshore removal and onshore disposal safely and in an environmentally responsible manner will continue to be priorities,' said spokesman Richard Webb.

Vocabulary-building activities

1

Making it happen

Complete the sentences below using one word from each box.

call	carry (× 2)	deal	hook	put
set	slow	sort	speed	tow	work

down	off (× 2)	out (× 4)
up (× 3)	with	on

1 We can only ________ the jacket ________ to sea when the weather is good.

2 Sometimes, we have to ________ ________ the work until the weather is better.

3 I told the operator to ________ ________. He was going far too fast.

4 I told the operator to ________ ________. He was going far too slowly.

5 We had to ________ ________ a lot of paperwork before the work could start. (two answers)

6 Before operations can begin, we have to ________ ________ and test all the equipment. (two answers)

7 The maintenance team needs to ________ ________ hundreds of different tests.

8 You're doing fine. If you ________ ________ like that, you will be very successful.

9 With a laser scanner, it is much easier to ________ ________ the dimensions of the pipes.

10 Unfortunately, we have had to ________ ________ the meeting. No one can make it at that time.

2 Putting up and taking down

Which of these verbs and verb phrases do you associate with the erection of a rig, and which do you associate with dismantling? Write the words in the correct columns.

abandoned   attached to the deck   broken up for scrap   decommissioned
disposed of or recycled   installed   lifted into place   mounted
detached from its mountings   secured to the seabed   towed back to shore   unscrewed

Erection	**Dismantling**

Refresh

Refresh the vocabulary you have learnt in this unit.

Put it to work

Have you experienced the erection or dismantling of a rig or other large plant? Describe what happened in as much detail as possible.

Section 2
Working on a Rig

8 Rig Automation

In all areas of oil and gas exploration and production, robotic automation is playing an increasingly important role. This unit focuses on vocabulary which will help you to discuss robotic automation and its impact on the industry in English.

Focused reading

At the 'Oil and Gas Futures' conference in Houston, Texas, many different presenters spoke about the products their companies are introducing to the industry. Here are two examples.

1

The way of the future
Anders Røyrøy is the Marketing Director of Hydro, a Norwegian company, which manufactures robotic systems for oil and gas extraction on topside platforms. Read the extracts from his presentation, below. Do you agree or disagree with his forecasts about the future?

- Technology will play a crucial role in increasing recovery and discovery of new fields. Currently, the biggest cost of oil production is people. In the future, once investment in technology has been made, recovery costs could be reduced dramatically.
- In the future, offshore platforms could be run by robots alone, with human beings staying on land. The petroleum operators will not find themselves in close contact with the new robots. The operators will remain on land and control them from there, reducing both risks and costs.
- The whole platform will be adapted for the robots. With an internal layout in the form of shelves, these platforms might look like hi-tech warehouses, with the robots moving up and down the rows of shelves like fork-lift trucks.

- Other robotic applications would include automatic tool changes, opening and closing of valves and inspection operations, such as measuring temperature, noise or vibration.
- Unlike a human being, a robot will not stop moving of its own accord or move aside if it collides with something else. On a platform where a number of robots are in operation, there could be collisions between them and other equipment. Operators onshore will have control systems to prevent collisions offshore.

2

New technologies

Lars Raunholt is CEO of Seabed Rig AS, a company which has developed an innovative drilling rig to be located on the seabed for cost-effective drilling in deep waters and arctic waters. Here he describes the rig, referring to the slide below.

What advantages of the seabed rig does he mention, as compared with conventional drilling methods?

Seabed Rig

As you can see in this slide, the seabed rig is connected to the surface vessel by a flexible umbilical, which transports power and data, as well as cement, mud, etc. There are no rigid risers and so it is far more flexible than conventional subsea systems. We have developed and patented a modular design for the rig, in which each component is housed independently and can be easily replaced. All drilling operations are carried out remotely, from the surface vessel, and there is a high level of redundancy to ensure that the rig is completely reliable. A key feature is a sophisticated mud pump system, specially adapted to seabed drilling. This is an environmentally friendly solution with no discharges and with better safety parameters than conventional drilling.

Vocabulary-building activities

1 Bullish, quietly confident or not so sure?

1 Underline the key phrases used for making forecasts. The first one is done for you.
- a Technology <u>is certainly going to</u> play a crucial role in increasing recovery.
- b In the future, offshore platforms could well be run by robots alone.
- c It is quite likely that operators will not need to be transported out to the rigs.
- d Recovery costs could easily be reduced by up to 50%.
- e These platforms might possibly look like hi-tech warehouses.
- f It is highly unlikely that supplies of oil and gas will increase.
- g We are fairly convinced that seabed rigs will replace topside platforms for subsea drilling.
- h The profitability of many oilfields will undoubtedly depend on the prices of oil and gas.
- i We are pretty sure that the oil and gas industry will continue to play a major role in the global economy.

2 Now identify which six forecasts are confident of the future and which three are not so confident.

2 What do *you* think?

Use the above phrases to make your own forecasts about the following statements.

1. The price of oil will rise above $200 per barrel.
2. The US government will limit the use of air conditioning in public buildings.
3. The airline industry will agree to cut carbon emissions by 50%.
4. Big oil and gas companies will have to pay a special 'environment tax'.
5. There will be a number of big takeovers or mergers in the oil and gas industry.

What would happen if these forecasts came true? Complete these sentences using your own ideas.

Example:

1. *If the price of oil rose above $200 per barrel, many transport companies would go out of business.*
2. If the US government limited the use of air conditioning in public buildings, ...
3. If the airline industry agreed to cut carbon emissions by 50%, ...
4. If big oil and gas companies had to pay a special 'environment tax', ...
5. If there were a number of big takeovers or mergers in the oil and gas industry, ...

3 Robotics

Use these words to complete the following passage about robotics in oil exploration and production.

actuators autonomously collision grips interface manipulates recognition redundancy remotely sensors signals unmanned

Robotics in Oil E&P

Robotics are increasingly used in oil exploration and production. Although not fully **(1)** ______________, as there will always be a need for computer operators, the technology does allow operations to be controlled **(2)** ______________, and therefore much more safely and efficiently. A robot is equipped with a number of **(3)** ______________, which give it a detailed awareness of its environment. **(4)** ______________ are sent back to the operator's computer, enabling instructions to be relayed in turn to the robot, which can then operate **(5)** ______________. The arm of the robot functions through a series of **(6)** ______________, which work in the same way as the muscles of a human arm. The mechanical hand is thus controlled as it **(7)** ______________ and **(8)** ______________ tools, valves, measuring devices, etc. To ensure reliability, there is always a high degree of **(9)** ______________ in a robotic system, meaning that if one operation fails, another is available to do the same job. Another issue which has to be overcome is **(10)** ______________ avoidance, as robots do not stop moving when something is in the way! The **(11)** ______________ between human operator and robot is usually in the form of computer commands, but increasingly it is being provided through more sophisticated methods, such as voice **(12)** ______________.

Put it to work

Either:
Describe a robotic or other unmanned exploration or drilling system with which you have worked, or which you have seen in operation. What are the advantages of this system compared to a more conventional manned operation?
Or:
What future do you predict for the oil and gas industry?

Refresh

Refresh the vocabulary you have learnt in this unit.

Glossary

Unit 7

Rig erection	
platform fabrication	building or construction of the platform
installation	putting a piece of equipment in its correct place and making it work
tow out to sea	pull away from the coast
up-end it into position	turn from its side and move it into the correct place
flood the ballast tanks	fill the tanks on the legs of the platform with water (ballast helps something to remain stable on the water)
secure it to the seabed	keep it in one place on the bottom of the sea
lift it into place	move a heavy object from one place to another
carry out the erection of the structure	be responsible for the building of the platform from start to finish
hook up and test the equipment	install the equipment and check that it works correctly
up and running	working normally after installation
mounted	attached to a flat surface
Rig dismantling	
dispose of	break up or sell something when you don't need it anymore
disposal	the act of disposing
decommission	when a rig is decommissioned, it is taken out of service
abandonment	the act of leaving a well after production is finished
cease production	stop production
debris	all the waste and destroyed material left after a rig has been decommissioned
broken up for scrap	scrap is metal which is no longer needed for its original purpose, but can be reused or recycled
detached from its mountings	separated from the mountings, which are what holds a piece of equipment to a machine ('detached' is the opposite of 'attached')
unscrew	to remove a piece of equipment which has screws

Unit 8

Platform/rig construction	
layout	the way the parts of a rig are organized in the space available
warehouse	a building used for storage in a factory or port, for example
seabed	at the bottom of the sea
umbilical	a line connecting two machines which allows a flow of material and/or information (a baby is connected to its mother by an umbilical cord)
riser	a rigid construction between the platform and the seabed
modular design	a design which allows pieces to be connected and separated independently
topside platforms	a platform above the surface of the water
Phrases used for making forecasts	
could well + verb	phrases used to make a confident forecast about the future
could easily + verb	
undoubtedly	
highly likely / unlikely to + verb	
might possibly + verb	phrases used to make a less confident forecast about the future
fairly convinced that	
pretty sure that	
quite likely to + verb	
Robotics	
unmanned	without human operators
controlled remotely	controlled from a distance
sensor	a device which detects (or 'senses') heat, noise, vibration, etc.
autonomous	independent
actuator	part of a robot's arm which enables the arm to move
grip	to hold in the hand
manipulate	to move with the hand
redundancy	when there is more than one machine or person to do the same job
collide with / collision	to crash into: a crash between two people or machines (e.g., cars)
interface	the connecting point between a person and a computer or between two computers
voice recognition	when a computer understands commands given by the human voice

Drilling Operations

	Titles	Focused Reading	Vocabulary Areas	Aim(s)
1	Drilling for Oil and Gas	• Online encyclopedia entry on the drilling process and a history of drilling	• Parts of the rig • Measurements • Approximations	To be able to describe the different parts of a rig, including measurements.
2	The Hoisting System on a Drilling Rig	• Article and photo from an engineering magazine • Supply document about Active Heave Drawworks • Diagram of a rotary rig hoisting system	• Components of the rig hoisting system • Drilling line operations	To be able to describe the operation of a hoisting system, with special focus on safety issues.
3	Power Systems	• Newspaper article about the upgrading of power systems • Sales brochure on drilling systems	• Components of the system • Upgrading / making improvements	To be able to describe the power system of a rig and its upgrading.
4	Rotating Systems	• Slide show presentation and transcripts about the basics of rotary drilling	• Parts of the rotating system • Operation of the system	To be able to describe the core parts of a rotary drilling system.
5	Circulating Systems	• Introductory explanation of the uses of mud • Sales literature about drilling fluids	• Components of the circulation system • Ingredients and types of mud • Functions and properties of mud	To be able to describe the circulation system of a rig.
6	Drilling Problems	• Journalist and platform manager interview transcript about drilling problems	• Reasons for downtime • Useful phrasal verbs • Prevention and solutions	To be able to describe drilling problems and know how to address them.
7	Preventing Blowouts	• Case study of a blowout • Transcript of a telephone call during an emergency	• Describing a blowout • Consequences • Describing a blowout preventer	To be able to describe what happens in the build-up to a blowout and its consequences.
8	Optimized Drilling	• Article from an engineering website about developments in drilling technologies	• New drilling technologies • Comparative vocabulary • Connecting words	To be able to describe new drilling technologies you have experienced.

Section 3
Drilling Operations

1 Drilling for Oil and Gas

Drilling operations are central to the oil and gas industry – both in exploration and production. The earliest oil wells were made by simply hammering at the earth. Developments in industrial engineering in more recent times mean modern wells drilled using rotary drills can achieve lengths of over 12,000 metres. This unit focuses on vocabulary which will help you to describe modern drilling operations in English.

Focused reading

1 Check the facts

Here are two extracts, opposite, from an online encyclopedia, which give an introduction to drilling.

1 Use the extracts to find the following information:
   a the minimum diameter of a drilling hole
   b the purpose of drilling mud
   c what happens to rock cuttings
   d the length of a joint of pipe
   e when petroleum was first used in the Middle East
   f the origin of the word 'petroleum'
   g the market share of the Baker-Hughes company

2 Complete the label of this photo using information from the extracts.

Oilfield in California, 1938. The first modern well was drilled around ____________ years earlier than this in Baku, in modern Azerbaijan.

http://www.encyclopediaonline.com

A summary of the drilling process

An oil well is created by drilling a hole 5 to 36 inches (127.0 mm to 914.4 mm) in diameter into the earth with a drilling rig, which rotates a drillstring with a bit attached.

These are the key steps involved in drilling a well:

- The drill bit, assisted by the weight of drillstring and drill collars above it, breaks up the earth.
- Drilling fluid ('mud') is pumped down the inside of the drillpipe and exits at the drill bit, helping to break up the rock, keeping pressure on top of the bit, as well as cleaning, cooling and lubricating the bit.
- The generated rock 'cuttings' are swept up by the drilling fluid as it circulates back to the surface outside the drillpipe.
- The cuttings and returning fluids are monitored for abnormalities to detect any possible 'kicks' in pressure.
- The pipe or drillstring to which the bit is attached is gradually lengthened as the well gets deeper by screwing in several 30-foot (10 m) joints of pipe at the surface.

This process is all facilitated by a drilling rig which contains all necessary equipment to circulate the drilling fluid, hoist and turn the pipe, control downhole pressures, remove cuttings from the drilling fluid and generate onsite power for these operations.

http://www.encyclopediaonline.com

Did you know?

- The first oil wells were drilled in China in the 4th century or earlier. They reached depths of up to 243 metres and were drilled using bits attached to bamboo poles.
- The Middle East's petroleum industry was established by the 8th century, when the streets of the newly constructed Baghdad were paved with tar, derived from petroleum that became accessible from natural fields in the region.
- The first modern oil well was drilled in 1848 by Russian engineer F. N. Semyonov, on the Aspheron Peninsula north-east of Baku. By 1861, Baku produced about 90% of the world's oil.
- The word 'petroleum' derives from two Greek words meaning 'rock oil' ('petro–leum').
- When Edwin Drake struck oil in the first US discovery in 1859, he was actually boring for salt.
- Until the 1970s, most oil wells were vertical. Using deviated and horizontal drilling, it has become possible to reach reservoirs several kilometres away from the drilling place.
- The original patent for the rotary rock bit was issued to Howard Hughes Senior for his dual cone roller bit in 1909. It consisted of two interlocking wheels.
- In 1933, two Hughes engineers invented the tricone bit. In 2000, the Baker-Hughes company still had a 40% share of the world market in drilling bits.

Vocabulary-building activities

1

Parts and systems

It is conventional to categorize the different parts of a drilling rig into four different systems: the power system, the hoisting system, the rotating system and the circulating system. In the diagram below, the prime movers are the basis of the power system. The other parts are the main components of the other three systems. Complete the table by putting the names of the different parts in the correct column.

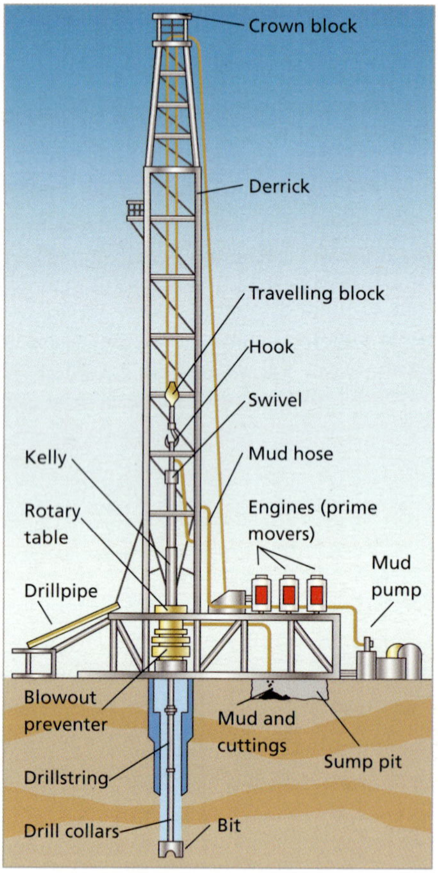

A typical rotary drilling rig

Hoisting system	Rotating system	Circulating system

2

Measurements

The following example shows two different ways of asking the same question:

A: How long is the pipeline?
B: What is the length of the pipeline?

The first uses the adjective 'long'; the second uses the noun 'length'. Complete the word pairs by filling in the blank spaces in the table. The first one is done for you. Use a dictionary to help you if necessary.

Adjective	Noun
long	length
wide	
	depth
high	
thick	
heavy	weight
old	
	price

3 Visitors' questions

A visitor to a drilling rig is asking the supervisor some questions about the rig. English is not his first language.

1 See if you can help him by completing his questions using words from the box. Start each question with 'What is the …?' or 'How …?'

diameter	long	deep	height	thick	cost	pressure	wind load

a Q: ______________________ is the well?

A: There are several wells on this rig. The deepest is around 4,500 metres.

b Q: ______________________ at that depth?

A: It can be as much as 15,000 psi, which is why we have a BOP to prevent blowouts.

c Q: ______________________ of the derrick?

A: It is approximately 40 metres from the crown block to the working floor.

d Q: ______________________ of the derrick?

A: It can withstand wind speeds of anything up to 125 miles per hour.

e Q: ______________________ of the hole?

A: The hole gets narrower as you go down, but at the surface it is more or less 50 cm.

f Q: ______________________ are the drillpipes?

A: These pipes are made of steel with a thickness of just under 1 cm.

g Q: ______________________ are the drillpipes?

A: Each joint is roughly 10 metres in length. If they are all the same, it makes storage much easier.

h Q: ______________________ of a diamond bit?

A: They are extremely expensive, ranging from $12,000 to $15,000 each.

2 Reread each of the answers and underline the phrases which indicate that the answer given is an approximation.

Refresh

Using one of the strategies for consolidating and storing new vocabulary recommended in the Introduction, refresh what you have learnt in this unit.

Put it to work

Sketch a diagram of a rig on which you have worked, or one you have visited. Label the different parts of the rig and try to use some of the vocabulary you have learnt to describe, for example, the height of the derrick, the depth of the well, the pressures, wind load, etc. Refer to information on the Internet to fill out your descriptions.

Section 3
Drilling Operations

2 The Hoisting System on a Drilling Rig

The most visible part of a drilling rig is the hoisting system, which comprises of a derrick, or mast, and a variety of steel ropes and pulleys for hoisting the drilling equipment into and out of the well. This unit focuses on vocabulary which will help you to describe the hoisting system in English.

Focused reading

1 Derrick or mast?

1 Read the passage from a manual for drilling trainees and study the picture. Does the photo show a derrick or a mast? How do you know?

> The derrick or mast is the steel tower directly above the well. A derrick is erected on site, piece by piece, whereas a mast arrives at the site on a tractor and trailer and is then jacked up into position. For example, a cantilever mast rig is assembled horizontally and then pivoted vertically, using the travelling block and drawworks of the rig.

2 Looking at the picture, can you describe the positions of the following after the mast has been erected:

- a the drawworks
- b the crown block
- c the travelling block

2 Engineering issues

Read this extract about drilling masts. It comes from the beginning of an article in a specialist engineering magazine. What issue do you think the article will go on to address in further detail?

Drilling Masts

A drilling mast is a thin, metallic structure that generally consists of tubular bars rigidly assembled together at their ends, called elastic nodes. The mast, erected above the opening of a well, enables the vertical motion of the travelling block and has the following major functional roles: (i) to sustain – by means of the hoisting system drawworks, drilling line, crown block, travelling block, hook and swivel – the drillstring, the casing string and the tubing string, and to make possible their up-and-down operation; (ii) to store the drillpipes, drill collars and tubing; and (iii) to facilitate the drilling and tool-running procedures. Consequently, the drilling mast is of vital importance to a drilling rig.

When in use, masts undergo strong vibrations, caused by the motion of the tubular string, of the driving equipment of the drilling machines, of the floating platforms (in an offshore set-up) and by winds. Excessive vibrations of these structures can generate undesirable effects and decrease significantly the lifetime of their parts.

3 Quantum-leap solutions

This is an extract from a document received by the purchasing department of an oil and gas company. Read the extract carefully.

1 What is the purpose of the document?
2 What is the most important benefit of AHD's improved active heave motion compensation?

NOV's Active Heave Drawworks (AHD) offer a quantum-leap solution to the offshore drilling industry by combining the latest technology in controls and design. The AHD has already set new standards for compensation, accuracy, improved wireline life, workable load range and rig safety by providing proven active heave motion compensation for floating rig applications. These drawworks eliminate the need for overhead motion compensation machinery, which provides a lower centre of gravity and less operational cost. The AC electric motors provide improved speed and torque control independent of load and speed, and they provide the ability to keep the maximum hook load at a standstill indefinitely. In addition, braking energy is regenerated and fed back into the drilling rig electrical system, which increases overall rig efficiency. And due to active heave compensation, the 'drilling operational window' is increased by allowing drilling programmes to continue in heavier seas than conventional drawworks.

Source: National Oilwell VARCO (www.nov.com)

Vocabulary-building activities

1

The rotary rig

Examine the diagram of a rotary rig hoisting system and complete the passage below.

The **(1)** ______________________ is made of braided steel wire about 3 cm in diameter. It is spooled around a revolving wheel on the drill floor called the **(2)** ______________________ and its speed is controlled by the driller using a brake lever. It goes over a pulley in the **(3)** ______________________ attached to the top of the mast and then down to the **(4)** ______________________ suspended in the derrick. There is a **(5)** ______________________ at the end of the line for attaching equipment.

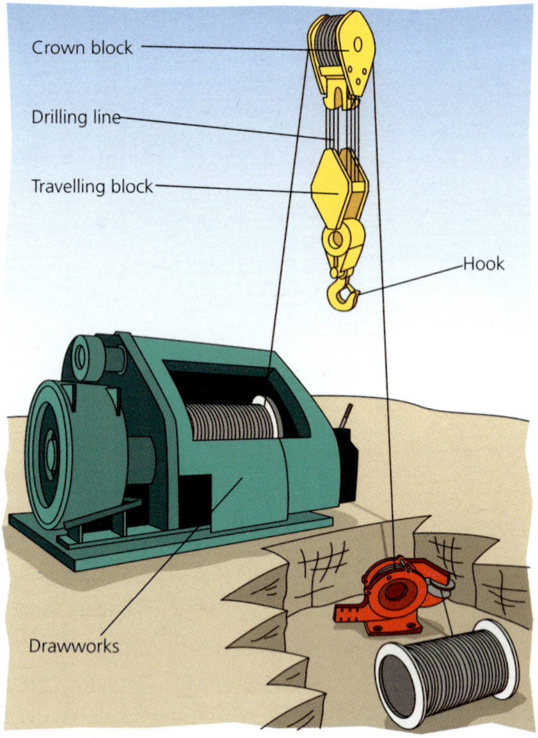

Rotary rig hoisting system (derrick not illustrated)

2

The drilling line

A new consignment of drilling line has arrived at a rig. The following sentences describe the different stages of the process of fitting the new line. Put each stage of the process into the correct order.

First of all, the end of the line was detached from the supply reel and attached to the crown block.

1. It was then threaded through one of the sheaves of the travelling block and back up to the crown block.
2. The line was then attached to the drum in the drawworks. This line is the fast line, the one that does the work of raising and lifting equipment.
3. The drilling line was threaded over one of the sheaves in the crown block and then lowered to the rig floor.
4. Meanwhile, the other end of the line was secured to an anchor mounted in the rig substructure. This line is the dead line, the one that is fastened to the rig.
5. This process was repeated until the correct number of lines had been strung up. We were using a ten-line string.

Finally, the travelling block was raised off the rig floor into the derrick, using the drawworks.

3 The drawworks

This is a detailed description of the drawworks on a rig. Match up the beginning and end of each sentence to form a complete description.

1	The drawworks is a big heavy piece of equipment	a	which slows down and stops the drilling line when necessary.
2	At the heart of the drawworks is a revolving drum	b	is the main shaft known as the catshaft.
3	Running through the centre of the drawworks	c	to help absorb momentum.
4	There are also a number of other shafts, clutches and drives	d	and are used to manoeuvre equipment around the rig.
5	It also contains a main brake and brake lever	e	one mechanical and one friction operated.
6	For especially heavy loads, there is also an auxiliary hydraulic and braking system	f	around which the drilling line is spooled.
7	On the catshaft there are normally two catheads mounted at each end,	g	the other for breaking out.
8	One of the mechanical catheads is used for making up drillpipe,	h	used for changing speed and direction.
9	The friction catheads are manually operated	i	powered by the prime movers on the rig.

Put it to work

You have been asked to deliver a training session to a group of junior field engineers. Your topic is the hoisting system of a rig. Prepare a short presentation for your trainees, using diagrams to show them the different parts of the hoisting system and how they work. Try to identify important elements of the design that aim to compensate for the stresses caused by difficult drilling conditions.

Refresh

Refresh the vocabulary you have learnt in this unit.

Glossary

Unit 1

Parts of the rig	
derrick	a steel tower on four legs which is raised in sections above the drilling floor
drilling line (or hoisting line)	steel cable used to raise and lower equipment downhole
drillstring	a combination of several different parts suspended from the drilling line downhole
crown block	a fixed steel frame with steel wheels on a horizontal shaft located at the top of the derrick
travelling block	a steel frame with steel wheels on a horizontal shaft suspended in the derrick by the drilling line
hook	attached to the drilling line, made of curved steel and used for suspending the swivel and drillstring in the well
swivel	a device which allows the drillstring to rotate while suspended from the derrick
kelly	a four- or six-side pipe located at the top of the drillstring
turntable	a rotary plate in the drilling floor which is powered by the prime movers and turns the drillstring (also called the rotary table)
drill collars	a heavy large diameter pipe run on the bottom of a drillstring to add weight to the bit
bit	the tool used for cutting the rock – a tricone bit, for example, is one with three cutting sections
mud	the chemical drilling fluid pumped downhole through the drillstring and out of the bit
mud hose	a rubber hose that connects the mud pump to the swivel
cuttings	the pieces of rock produced by drilling
blowout preventer (BOP stack)	a series of devices attached to the top of a well to close the well as necessary during drilling
prime movers	diesel engines that supply power on a rig
Measurements	
length *(n)*	horizontal measurement from one end to another (long, *adj*)
height *(n)*	vertical measurement of an object (high, *adj*)
thickness *(n)*	measurement through an object (thick, *adj*)
weight *(n)*	measurement of heaviness of an object (heavy, *adj*)
width *(n)*	horizontal measurement of an object from one side to another (wide, *adj*)
pressure	the force caused by pushing, measured by the amount of force exerted on a particular area
wind load	the speed of wind that can be resisted by an object
narrow	opposite of wide
roughly	word indicating that the figure used is not exact – either a little more or a little less than the exact figure

Unit 2

Components of the rig hoisting system	
mast	a steel tower that can be constructed off-site and then raised into position on-site
cantilevered	with a mechanism allowing something to be raised and lowered
drawworks	a drum in a steel frame driven by engines (or 'prime movers'), used for raising and lowering equipment downhole via a drilling line wound around a reel in the drum
nodes	connecting points between the bars on a derrick or mast
braided	twisted together to strengthen (rope or cable)
pulley	a wheel used in the hoisting system – the line passes around the pulley to enable it to move freely
supply reel	the drilling line is purchased on a reel
sheaves	in the crown block and travelling block there are a number of sheaves containing pulleys through which the drilling line travels
drum	the drilling line is spooled around the drum – the central part of the drawworks
fast line	the moving line between the drawworks and the travelling block
dead line	the non-moving line from the crown block, attached to the rig floor
substructure	the structure of the drilling rig below the drilling floor, but above the surface of the hole
shaft	a rod or bar in an engine that turns around as the engine works
catshaft	the main shaft in the drawworks
absorb	to take in – for example, to absorb pressure or information
clutch	the mechanism in an engine which allows the engine to move into a different gear
drive	the part of the engine which supplies power to its moving parts
brake	the mechanism for slowing down and stopping an engine
brake lever	a handle used to manually control a brake
manoeuvre	to move about from one place to another
cathead	the end of the catshaft which protrudes from the side of the drawworks
Drilling line operations	
hoisted	pulled up and down
threaded (or reeved)	pushed through a hole or a narrow space
fastened (or secured)	tied to a fixed structure
spooled (or wound)	tied around a drum or spool
attached	connected to a fixed structure
detached	disconnected from a fixed structure
secured to an anchor	tied to a heavy weight or structure to stop it moving

Section 3
Drilling Operations

3 Power Systems

With developments in electronics, the technology of the power system on a drilling rig has been transformed in recent decades from the noisy and polluting diesel engines of the early years to the quiet, smooth-running motors of today's rigs. This unit focuses on vocabulary which will help you to describe power systems in English.

Focused reading

1 Upgrades

Read this article about upgrades to TNK-BP rigs. Why have they made these changes?

http://www.abb.com

Oil giant improves rig performance with ABB drilling drives

TNK-BP, one of the world's largest oil producers, is carrying out a $600-million modernization programme to improve the safety and performance of its oilfield drilling rigs. At the heart of the rig revamps is an ABB innovation – AC drilling drive systems.

The objective of the modernization programme is to improve rig safety, extend rig operating life, increase drilling efficiency and raise cost-effectiveness. The target is to achieve performance improvements of between 10 and 30%. 'AC systems are lighter, smaller and about 25% more efficient than other systems. They allow faster, safer rig moves and have less potential for human error in rig up,' said a TNK-BP spokesman.

About 40 of TNK-BP's fleet of 100 drilling and sidetracking rigs are slated for improvement. The scope of the revamps is comprehensive and ranges from enclosed operator control cabins to new top drives, hoisting systems and mud pumps. At the heart of the revamp concept is an AC drilling drive system – a pioneering ABB innovation from the 1990s that enables the rigs to be powered by cutting-edge AC drives and motors, rather than conventional and less advanced DC powered systems.

Source: Retrieved from www.abb.com/cawp/seitp202/95c017d775c2d2fdc12573b60035494a.aspx

2

Key selling points

The following extract from an ABB sales brochure gives details of the ABB AC drilling drive system and drilling motors. Read the extract and underline the key selling points of the system.

ABB AC Drilling Drive System

Our AC Drilling Drive system consists of robust and virtually maintenance-free induction motors controlled by a modular, solid state water-cooled frequency converter system. The motors need no pressurizing and air cooling eliminates pipe systems or heat exchangers. They can operate without an eddy current brake system, reducing weight and saving space. The common DC bus principle of the converter system means one supply section, including cabling cost and space requirements. Expansion is accomplished by connecting new converters to the DC bus.

AMA Drilling Motor

ABB AC Drilling Motors

A major advantage of an AC drive is the use of an induction motor. This is the most versatile, reliable and economical type of motor. Our motors are standard frame-size motors, specially designed for frequency control and adapted to drilling applications.

- High efficiency, low noise/vibration
- Simple construction, less maintenance
- Constant torque in full speed range
- Frame size: 350–400–450–500–560
- Output up to 4300 kW, 600/660/690 V
- Separate mounted ventilator fan, air cooled (no water cooling or air-purging system required)
- Water cooling / air cooling
- ATEX certified

Source: Retrieved from www.library.abb.com/global/scot/scot201.nsf/veritydisplay/9f4f8b1d51b7b00fc2256f7b002aa463/$file/ac_drilling_brochure_low_resolution.pdf

Vocabulary-building activities

1

Powering the rig

The diagram below shows the typical layout of a modern drilling rig power system.

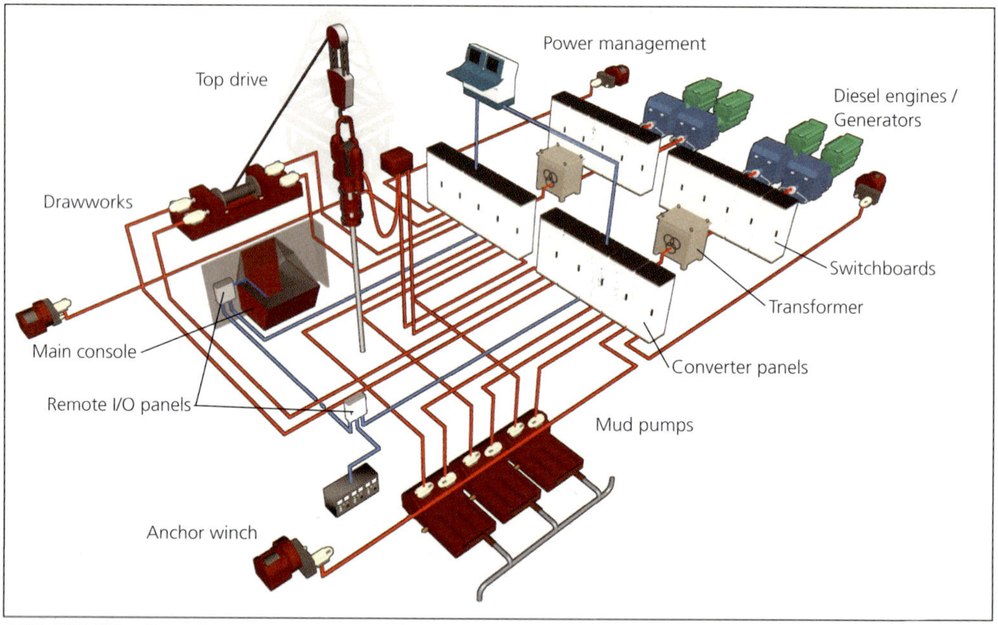

ABB AC Drilling Drive System

Which of the items identified on the diagram ...

1 changes electricity from direct current (DC) to alternating current (AC)?
2 creates electricity?
3 ensures that there is a reliable, uninterrupted supply of power?
4 ensures that current is at the correct frequency and directs the current to the different items on the rig?
5 receives and then redirects the electric current using a series of switches and fuses?
6 provides an area for operators to monitor and control the entire drive system?
7 powers the drillstring during drilling operations?

2 System upgrades

The development of top drive drilling and, more recently, the development of AC drilling motors to power top drives, has been essential in the modernization of oil and gas drilling. Here are some of the positive changes which have occurred as a result of these innovations.

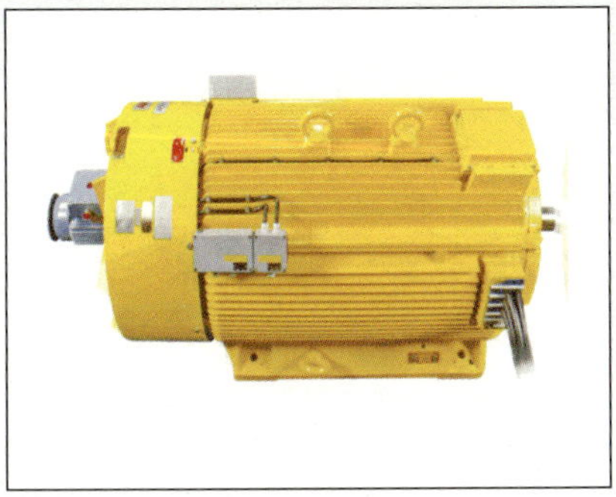

HXR Top Drive Motor

Top drive drilling system

Use your own knowledge and experience to complete the sentences below using comparative adjectives (for example, 'bigger', 'more expensive', etc., or just 'more' or 'less' on their own). Your completed sentences should describe positive changes that have resulted from modernizations in drilling techniques. You can use these words to help you.

compact	efficient	great	less	long	low	more	profitable	reliable	simple

1 The costs and risks associated with ultra-deep and extended reach drilling are much ________.
2 As a result, recovery from marginal fields has become ________.
3 The speed and torque control of top drives is ________.
4 Modular drilling systems have been developed, allowing for a much ________ system and ________ time for installation.
5 There is ________ redundancy, which means ________ downtime if things go wrong.
6 Operation and maintenance is ________ and ________.
7 Equipment has become ________ and lasts ________.

Put it to work

Imagine you have to put the case to your manager for upgrading the power system in operation on a rig on which you have worked. Describe the power systems in operation, the ways in which they could be improved upon and the benefits that would result.

Refresh

Refresh the vocabulary you have learnt in this unit.

Section 3
Drilling Operations

4 Rotating Systems

The rotating system is the business end of the drilling rig. Supported by the hoisting system and driven by the power system, it is the rotating system that actually makes the hole in the rock formation. This unit focuses on vocabulary which will help you to describe the rotating system in English.

Focused reading

1 Slide presentation

Anwar is training a group of field service engineers and he is giving them a presentation about the basics of rotary drilling using a series of PowerPoint slides. The transcript of what he says about each slide is on the opposite page.

1 Notice how the slides, below, are used as a basis for Anwar's presentation. Read the transcripts and then number the slides on the lines provided.

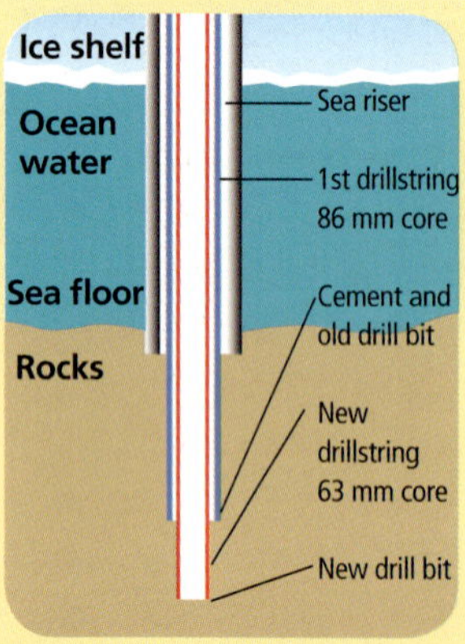

a ________________

Source: www.drillers.com/pages/view/photo-gallery/

b ________________

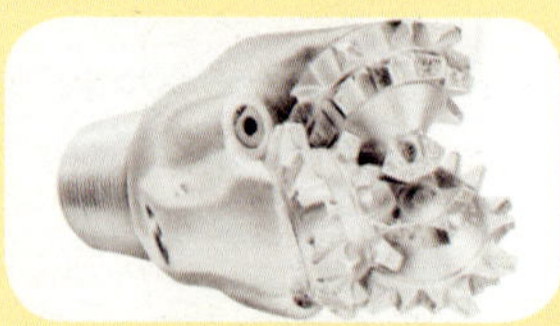

c ________________

d ________________

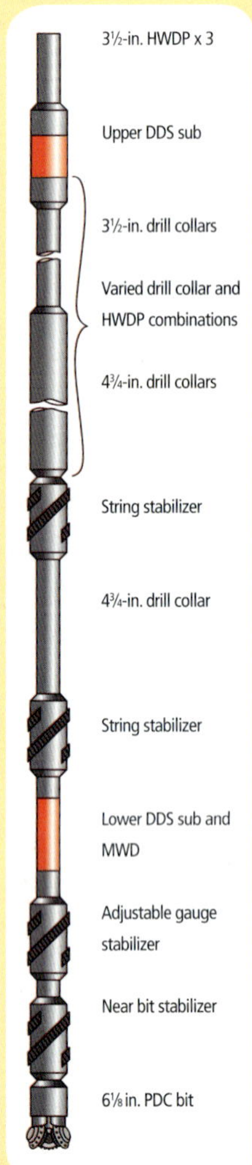

e ________________

Slide 1

If you look at this first slide, you can see that often there is more than one string. This diagram shows an operation in Alaska where a riser has been passed through the ice and down to the seabed, and the cement casing is in place in the rock. Only the innermost string is actually operational.

Slide 2

The bottomhole assembly (BHA) is made up of a drill bit which is used to break up the rock formations, drill collars (which are heavy, thick-walled tubulars used to apply weight to the drill bit) and subs (smaller sections of pipe), such as stabilizers, which keep the drilling assembly centred in the hole. The BHA may also contain other components, such as a downhole motor, Rotary Steerable System, measurement while drilling (MWD) and logging while drilling (LWD) tools.

Slide 3

Heavyweight drillpipe (HWDP) is used to make the transition between the drill collars and drillpipe. The function of the HWDP is to provide a flexible transition between the drill collars and the drillpipe. This helps to reduce the number of fatigue failures seen directly above the BHA. A secondary use of HWDP is to add additional weight to the drill bit. Drillpipe makes up the majority of a drillstring. A drillstring is typically about 15,000 feet in length for an oil or gas well vertically drilled onshore in the United States, and may extend to over 30,000 feet for an offshore deviated well. The components of the string are joined together with special threaded connections known as tool joints, on the ends of each joint.

Slide 4

Drill tubulars are manufactured in 31-foot lengths, although they can also be manufactured in 45-foot lengths. Each 31-foot component is referred to as a joint. Typically, two, three or four joints are joined together to make a stand. Pulling the drillstring out of or running the drillstring into the hole is referred to as tripping. Drillpipe, HWDP and collars are typically tripped in stands to save time. This photo shows a device used for safe automated handling of double or triple stands.

Slide 5

This photo shows a typical tricone drilling bit. In today's modern industry, the two main types of drill bits are classed as PDC (Polycrystalline Diamond Compact) and Roller Cone; although the tricone dominates, bicone and mono cone bits do exist. Natural and synthetic diamonds are used in coring bits, as well as for very hard rock drilling with mud motors and turbines.

Vocabulary-building activities

1

Name the equipment

Identify the following items of equipment mentioned in Anwar's presentation in Focused reading 1.

1 Threaded connections used for joining together the different parts of the string.
2 An outer layer of piping thick enough to withstand the pressure of the sea, down which circulation fluids can be passed.
3 A steel and concrete shoring used in a formation to protect the walls of the well.
4 The part of the bottomhole assembly used to break up rock.
5 Heavy pieces of piping which add extra weight to the bottomhole assembly.
6 Devices that enable the drillers to keep the drillstring centred in the hole.
7 The combination of all the elements used downhole in the drilling process.
8 A length of pipe, typically 31 feet long.
9 A combination of pipes and other components, connected to the string above the rig floor as the string is extended.
10 A bit with three rotating parts for cutting and breaking the rock.
11 Another name for drillpipe, so-called because it is shaped like a tube.
12 These are smaller sections run between and below drill collars with various functions in the drilling process.

2

Fill in the blanks

Use the verbs given in the box below to complete the information about drilling bits. Use the 's' ending if you think it is necessary.

create(s) crush(es) decreas(es) drill(s) flow(s) make(s) rotate(s)
screw(s) turn(s) wear(s) out

1 The bit ________________ into the bit sub on the bottom of the drillstring.
2 Drilling mud ________________ through channels in the bit.
3 In a rotary cone bit, the cones ________________ on sealed and self-lubricating bearings.
4 The teeth or buttons on the bit ________________ the rock at the bottom of the well.
5 A PDC bit is the most expensive type, but it ________________ more footage in a well than any other type.
6 A bit ________________ at a rate of up to 200 rpm.
7 The bit ________________ cuttings which are washed back up to the surface by the drilling mud.
8 A tricone bit ________________ and needs to be replaced after 8 to 200 hours of drilling.
9 When a bit has worn out, rate of penetration ________________ and the drillpipe ________________ a different noise on the drilling floor.

3 Rotation

One of Anwar's students asks him a question: Exactly how does the drillstring rotate? This is his answer:

Thanks a lot Dieter, that's a good question. First of all, I will explain the traditional rotating system and then we can look at some of the latest technological developments. As you know, there is a *hook* attached to the hoisting line below the travelling block. Well, attached to the hook is a *swivel*, which in turn connects to the drillstring. Below that is a heavy four or six-sided pipe called the *kelly*, which is gripped by something known as the *kelly bushing*, which in turn fits into the *master bushing* of the *rotary table*. The kelly is connected to the drillstring, and so when the rig is powered up, all the components rotate together as a unit. Is that clear? Good … now let's look at some of the recent developments in power systems …

Use his answer to label the diagram below. The features to label are given in the paragraph above, in italics.

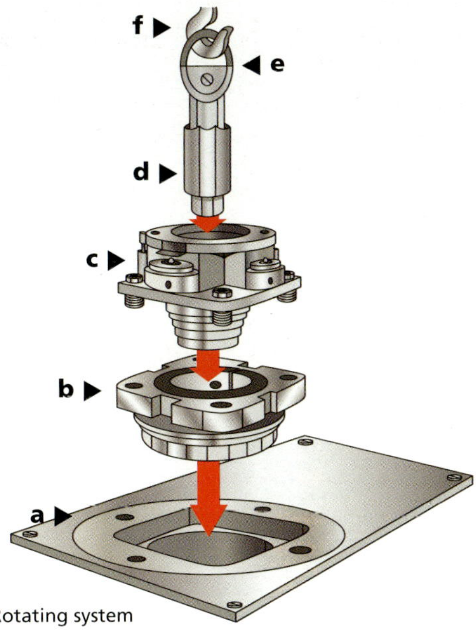

Rotating system

a ______________________

b ______________________

c ______________________

d ______________________

e ______________________

f ______________________

Refresh

Refresh the vocabulary you have learnt in this unit.

Put it to work

Imagine you are standing on the drilling floor of a rig with someone who is visiting the rig for the first time. Look around you and describe what you can see in as much detail as possible.

Glossary

Unit 3

Components of the power system of a rig	
current (DC/AC)	the flow of electricity, which can be direct or alternating
voltage	measurement of electromotive force or electric potential
transformer	a device which changes current from one voltage to another
fuse	a safety device which melts if overloaded to protect an electrical circuit
switchboard	a panel where flows of electricity can be controlled
console	an area where operators can safely control the power systems
top drive	a drilling device which allows drilling without the need for a rotary table
induction motor	a type of AC motor where power is supplied to the rotating device by means of an electromagnetic force (induction)
converter	a device for changing frequency of current
eddy current brake system	a system for slowing down and stopping a machine (brake), operated by means of electromagnetic forces which move in circular directions, like water around a rock in a river (eddy)
DC bus	a high voltage connection for conducting direct current, with separate connections
ventilator fan	a device for air cooling a machine
torque	a force that causes rotation
air-purging system	a system for removing air
Making improvements	
upgrade / modernize / revamp	to make improvements in order to make more efficient or effective
achieve performance improvements	to become more efficient in performance
slated for improvement	identified as needing improvements
a pioneering innovation	a new technique or technology which is the first in its field
cutting edge / leading edge / forefront	in first place in relation to technological competition
Selling points of a machine	
versatile	flexible – can be used for many different purposes
reliable	doesn't often break down
economical	cheap to use
compact	small, with an efficient use of space
redundancy	alternative ways of doing the same thing in case of failure of one part or system
reduced downtime	more time spent in operation – downtime is when the machine is not working

Unit 4

Components of the rotary system of a rig	
drillstring	a combination of several different parts suspended from the drilling line downhole
riser	a vertical pipe for the upflow of oil or gas
cement casing	a steel pipe lined with cement placed in a drilled hole to prevent the hole from collapsing
innermost string	where there are a number of strings inside one another, the drillstring nearest the centre of the hole
drill collars	a heavy steel tube, used between the drillstring and the bit to add weight to the bit
tubular	another name for a drillpipe
sub	a short piece of tube used to connect tubes which cannot be directly connected, often serving a specific function, such as for measurement or logging while drilling
stabilizer	a type of sub used to ensure that the drillstring remains centralized
threaded connection	a helical ridge on the end of a pipe which allows it to be joined to another pipe
tool joint	the name given to the threaded connection on a drillpipe
joint	a section of drillpipe
stand	a number of connected joints
bottomhole assembly (BHA)	the various pieces of equipment, including the drill bit and related tools in the lowest part of the drillstring
bearings	the parts of a machine that allow two or more parts to rotate or move in contact with each other with as little friction as possible
kelly bushing	a device which is connected to the kelly to create torque, allowing the kelly (and the rest of the drillstring) to be raised and lowered during downhole operations
master bushing	part of the rotary table which connects to the kelly bushing
The operation of the rotary system	
apply weight	add additional weight
reduce fatigue failure	lower the chances of failure because of excessive use
trip / make a trip	to lower and then raise the drillstring into and then out of the hole
coring	taking a sample of formation using a special tool (a coring bit)
withstand pressure	to resist pressure
shoring	preventing from collapse
crush	apply a heavy weight and break into many pieces
screw	to rotate, for example, into the formation
wear out	to fail after time because of use
wash back	to return to the surface in a fluid, such as drilling mud
grip	to hold tight

Section 3
Drilling Operations

5 Circulating Systems

During drilling, a multi-purpose fluid known as 'mud' is circulated in the borehole. Large quantities are required for the circulating system, and there are strict rules governing the chemicals which can be used because of potential environmental impacts. This unit focuses on vocabulary which will help you to describe the circulating system in English.

Focused reading

1 Multi-purpose mud

Read this introduction to the uses of mud on a drilling rig and underline the two or three words that best summarize the main point of each use.

Drilling fluid

Mud injection pumps

Drilling fluid (often called 'mud') is used to:

1 lift soil/rock cuttings from the bottom of the borehole and carry them to a settling pit
2 allow cuttings to drop out in the mud pit so that they are not recirculated (influenced by mud thickness, flow rate in the settling pits and shape/size of the pits)
3 prevent cuttings from rapidly settling while another length of drillpipe is being added (if cuttings drop too fast, they can build up on top of the bit and seize it in the hole)
4 create a film of small particles on the borehole wall to prevent caving and to ensure that the upward-flowing stream of drilling fluid does not erode the formation
5 seal the borehole wall to reduce fluid loss (minimizing volumes of drilling fluid is especially important in dry areas where water must be carried from far away)
6 keep underground pressures in check by adding weighting materials, such as barite, to the fluid
7 cool and clean the drill bit
8 lubricate the bit, bearings, mud pump and drillpipe

2 Types of drilling fluid

Baker Hughes is a major player in the market for drilling fluids. Read these extracts from their sales literature. Why does Baker Hughes offer three different types of drilling fluid?

Water-Based Fluids

Baker Hughes Drilling Fluids' high performance water-based muds (HPWBM) are designed to emulate the performance attributes of invert-emulsion systems. The need for environmentally safe and technically equivalent, water-based alternatives to emulsion systems is increasingly becoming an important consideration in the drilling fluid selection process. Baker Hughes Drilling Fluids was the first service company to introduce HPWBM, starting with the AQUA-DRILL System℠, in the early 1990s.

We have continually researched technology to constantly improve high performance water-based muds. New technology such as MAX-SHIELD℠ and MAX-PLEX℠, which improves shale stability, and MAX-GUARD℠, which suppresses the hydration and dispersion of heavy and reactive clays, have proven to be significant performance-enhancement products in our high performance water-based product line.

Specialty Fluids

As the drilling industry matures and the search for hydrocarbons moves into even more challenging environments, the need for innovative and novel drilling solutions takes on greater importance. The utilization of specialized drilling fluid additives is becoming more prevalent, especially in depleted fields and in deep and ultra deep waters. These challenging wells require more than just traditional additives to successfully achieve the goal of moving oil and gas from the reservoir into the refining process.

Baker Hughes Drilling Fluids offers a complete line of innovative specialty products for every drilling need. From basic components for controlling lost circulation, maintaining shale control and providing downhole lubricants, to the more exotic products designed to prevent differential sticking in demanding wells, no problem is too large or too small.

BHDF's product portfolio includes lost circulation materials, lubricants, ROP enhancers, sophisticated shale control products, and most importantly, spotting fluids. Since its first successful application in the 1940s, Baker Hughes Drilling Fluids' BLACK MAGIC™ spotting fluid has been widely recognized as the drilling industry's best product for successfully freeing differential stuck pipe.

Emulsion Fluids

Baker Hughes Drilling Fluids is on the forefront of developing synthetic-based compliant emulsion systems for drilling in challenging downhole conditions, such as deep-water and HPHT (high pressure – high temperature). The developments were driven by the extreme HPHT environment associated with Deep Shelf drilling in the Gulf of Mexico and the deviated HPHT wells drilled in the North Sea. Laboratory testing demonstrates fluid stability at temperatures in excess of 580°F and pressures exceeding 30,000 psi.

Since 2001, Baker Hughes Drilling Fluids has been working on perfecting a major change in HPHT emulsion-based systems. Although the revision was initially triggered by environmental legislation changes in the North Sea area, the developments and improved chemistries have demonstrated superior performance in both diesel-based fluids and in the synthetic fluids used in the Gulf of Mexico (GOM) and other areas. The chemistries chosen deliver minimal impact on the environment and to the health and safety of the personnel involved.

Source: Baker Hughes Incorporated (www.bakerhughesdirect.com)

Vocabulary-building activities

1

Identify the parts

Read this introduction to the circulation system of a drilling rig.

1 Using the vocabulary in the passage below, identify and label the five items which are missing on the diagram.

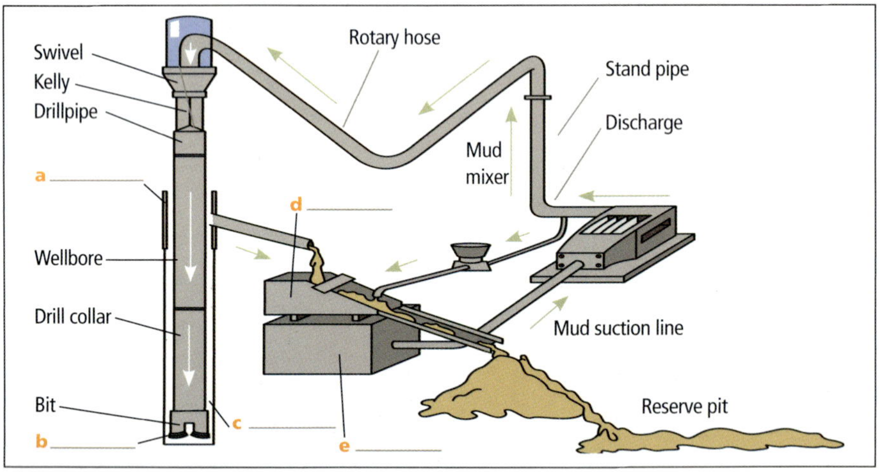

Mud circulating system

On a drilling rig, mud is pumped from the mud pits through the drillstring where it sprays out of nozzles on the drill bit, cleaning and cooling the drill bit in the process. The mud then carries the crushed rock ('cuttings') up the annular space ('annulus') between the drillstring and the sides of the hole being drilled, up through the surface casing, and emerges back at the surface. Cuttings are then filtered out at the shale shakers and the mud returns to the mud pits. The returning mud can contain natural gases or other flammable materials. These can collect in and around the shale shakers area or in other work areas, so safety procedures have to be in place and special monitoring sensors and explosion-proof certified equipment have to be installed. The mud is pumped back downhole and is continuously recirculated. After testing, the mud is treated periodically in the mud pits to give it properties that optimize and improve drilling efficiency.

2 Decide if the following statements are true (T) or false (F). If the statement is false, see if you can correct it.

		T	F
a	Mud travels from the mud pit, down the drillstring and out of the drill bit.		
b	The mud carries the cuttings back up the drillstring to the surface.		
c	The shale shakers are used to mix the mud.		
d	Mud can be dangerous after it has been downhole.		
e	Dangerous mud is disposed of.		
f	Drilling efficiency can be enhanced by adding supplementary fluids.		

2 Function, property or ingredient?

The following phrases in the box refer to either:

- the *functions* of mud (i.e., what it does during the drilling process)
- the *properties* of mud (i.e., why it is used for these functions and how it can be altered to enhance its functionality)
- the *ingredients* of mud (i.e., what it contains)

Put each phrase in the correct column in the table below.

clays, such as bentonite   density and gel strength   detergents and thinners to improve efficiency   improvement of ROP (rate of penetration)   lubrication of drilling tools   lubricity and viscosity   maintenance of pressure   protection of the borehole   reduction of fluid loss during drilling   solubility in water   suspension of cuttings   toxicity and danger to the environment   various polymers for specific purposes   water or a synthetic alternative

Functions	Properties	Ingredients

3 Complete the sentence

Use two of the phrases above to complete each of the following sentences. You may need to add one or two words to create correct sentences.

1 The basic ingredients of mud are …
2 Two of the most important functions of mud are the …
3 Additives are used in mud to modify its …

Put it to work

You have been given the responsibility of researching the different types of mud available in the market and of presenting your recommendation to senior management. Which type of mud would you use and why? Explain your choice with reference to the diagram of the drilling rig's circulation system on the previous page.

Refresh

Refresh the vocabulary you have learnt in this unit.

Section 3
Drilling Operations

6 Drilling Problems

Any type of problem that occurs during drilling can lead to downtime which, of course, is expensive and affects the profitability of an operation. Some worst-case scenarios, such as blowouts, can cause injuries to employees, or even fatalities, if action is not taken swiftly and efficiently to address them. This unit focuses on vocabulary which will help you to describe drilling problems and how to address them in English.

Focused reading

Alan Noughty, the journalist we met in Section 2, Unit 7, who works for a local newspaper in Aberdeen, is doing research for an article about developments in the oil and gas industry. He is with Jo Mcleod, a platform manager, who is giving him a tour of the rig and showing him some of the day-to-day operations.

1 Downtime

Alan records his interview with Jo so that he can listen again later. Read the transcript of their conversation as they approach the drilling floor. What is the problem? How is the problem being solved?

AN: There doesn't seem to be a lot going on here today. The rotary table isn't going round and there's nothing on the kelly.
JM: No, you're right, we've tripped out and put the kelly into the rathole for the time being. You can see the drillpipes are stacked over there in the rack.
AN: Why? What's the problem?
JM: There's a fish down the hole.
AN: A fish? How did that get there?
JM: I don't mean a real fish. It's what we call stuff that gets lost downhole. We also call it junk.
AN: So what have you lost today?
JM: It's not exactly lost, it's at the bottom of the hole. It's one of the bit cones. So we have to suspend drilling and bring in a fishing services company to retrieve it.

Fishing services team at work

AN: How long will that take?

JM: Well, hopefully not too long. They're very expensive, and every minute we're not drilling is time and money wasted. Once they've arrived and set up, it doesn't usually take too long to latch onto a fish. They're due to turn up in about thirty minutes or so. All being well, we'll be up and running again sometime tomorrow.

2

Troubleshooting

The conversation moves on to discuss some of the other problems that can occur during drilling. Make a list of the problems they discuss. How are these problems solved?

AN: So, what other problems can occur while you are drilling?

JM: Well, fortunately, technology has improved so much in the last few years that the number of incidents is declining all the time. Nevertheless, the pressure to increase drilling speeds does mean that the chance of making mistakes has gone up. The most dangerous and potentially expensive problem is a blowout. These are extremely rare, thank goodness.

AN: What is the most common?

JM: I suppose fish are the biggest problem, but we also get stuck pipe from time to time.

AN: What happens then?

JM: You mean how does it happen? Or what do we do about it?

AN: Well, both, I guess.

JM: Well, it can happen when the drillpipe gets literally stuck to the walls of the well. We call that differential pipe sticking, and we can usually free the pipe by jarring it, using a sub in the drillstring called, funnily enough, a jar. Sometimes we need to use a spotting fluid in the area where the sticking happens. Other times you get a deviation in the well – we call it a dogleg – which causes the drillpipe to stick. We can free it by reshaping the hole using a piece of kit called a reamer.

AN: I see. Anything else?

JM: Let me see. You can have different types of damage to the formation. Sometimes this means lost circulation, where the mud flows into the formation, for example. And you can also get sloughing.

AN: Sloughing? What's that?

JM: That's when the formation collapses.

AN: And how can you prevent these problems?

JM: Well, in most cases, the mud is the key. You can use particular types of mud; you can thin it down, thicken it, add different substances to it, and so on.

AN: So it's pretty complicated.

JM: That's right.

Vocabulary-building activities

1

Plug the holes

The sentences below describe some of the problems that may be encountered when drilling. Complete the sentences using one word from each box.

adhere	escapes	falls	forced
forces	got rid	lead	pass

into (x 2)	of	through
to (x 2)	up	from

1 A fish is a piece of drillstring, part of the bottomhole assembly or a tool that ________ ________ the hole.

2 Stuck pipe is when suction causes the pipe to ________ ________ the wall of the well or the hole is too narrow to allow the drillstring to ________ ________.

3 Sloughing shale is material from the formation that forms large balls which cannot easily be ________ ________ by the mud.

4 Formation damage occurs when the drilling fluid is ________ ________ the formation.

5 Lost circulation occurs when the formation is porous or highly fractured and an amount of mud ________ ________ the well.

6 Corrosive gases, such as carbon dioxide (CO_2) or hydrogen sulphide (H_2S), can ________ ________ corrosion and embrittlement of a steel drillstring.

7 Kicks occur when an abnormally high pressure below the surface ________ an uncontrolled flow of fluids ________ the well.

2

Everyday speech

Phrasal verbs are very common in informal speech. There are several examples in the transcript of Alan and Jo's interview (see Focused reading 1 and 2). Underline the phrasal verbs in the interview and identify their meanings in the table below, using the context of the interview to help you.

bring in   do about   go on   go up   set up   thin down   trip out   turn up

Phrasal verb	Meaning
	happen / take place
	pull the drillstring out of the hole
	hire / recruit / use
	arrive
	prepare everything to start work
	increase
	do something in order to solve a problem
	make less thick by adding extra water or other fluid

3 Preventing and anticipating problems

The best thing, of course, is to prevent problems before they occur. Do the following sentences describe preventions or solutions to drilling problems? Place a 'P' or an 'S' in each box to indicate 'prevention' or 'solution'.

1 Chemicals, such as potassium salts, are added to the drilling mud to inhibit sloughing shale. ☐

2 A lost circulation additive, such as fibre, flake or grain, is used when a zone of lost circulation is encountered. ☐

3 A stuck pipe can be freed by jarring the drillstring. ☐

4 Spiral-grooved drill collars are often used to prevent wall sticking. ☐

5 Formation damage in a well can be treated by well stimulation, acidizing or hydraulic fracturing. ☐

6 If you drill the formation using brine (very salty water), formation damage can be reduced. ☐

7 To reduce the risk of corrosion, it is necessary to use more expensive and resistant steel in the drillstring. ☐

8 To retrieve pipe from a well, either a spear or an overshoot is used, screwed into the bottom of a fishing string. (See the below pictures for other examples of fishing tools.) ☐

9 There are various methods of killing a well when a kick is detected, and hopefully it will not turn into a full-scale blowout. ☐

Refresh

Refresh the vocabulary you have learnt in this unit.

Put it to work

What drilling problems have you encountered? How were the problems solved? What could have been done to prevent them?

Glossary

Unit 5

Components of the circulation system	
mud	the drilling fluid used in the circulation system
mud pits	a storage area for mud before it goes into the system
settling pits	an area where cuttings are put after they have been extracted from circulated mud
shale shaker	the machine used to separate cuttings from circulated mud
rotary hose	the hose which connects the mud pump to the drillstring for discharging the mud
nozzles	holes in the drill bit, from which mud is sprayed
annular space (or annulus)	the space between the drillstring and the borehole spacing
Ingredients and types of mud	
emulsion fluids	drilling fluids containing oil
invert-emulsion system	an emulsion-based fluid which produces two layers, an inner layer where the cuttings are suspended and an outer layer which seals the borehole
clay	a sticky, heavy soil with very fine particles
additives	chemicals which are added to the fluid to perform a specific function
spotting fluids	a fluid used to deal with a specific area of the borehole, for example, where pipe is stuck
detergents	an agent which makes particles of soil, etc., more soluble and so easier to transport
thinners	a substance used to reduce the viscosity of the fluid
Functions of mud	
lift/suspend cuttings	hold cuttings and carry them up to the surface (lifting and suspension)
create a film	make a thin layer of mud on the inner surface of the borehole
seal the borehole	ensure that fluids cannot pass into the rock formation around the borehole
lubricate the bit	ensure that the bit moves freely at all times
enhance performance	improve performance
maintain shale stability	ensure that the rock, clay and mud (or 'shale') of the formation remains in place
prevent caving	to stop the borehole from collapsing
prevent the bit from seizing	to stop the bit from becoming jammed – for example, an engine seizes when it doesn't have any oil in it
hydrate and disperse clays	to add water and then separate the clays (hydration and dispension)
Properties of mud	
lubricity	the ability of the fluid to lubricate drilling equipment
viscosity	the thickness of the fluid – a viscous fluid is one which is semi-liquid, semi-solid
solubility	the ability of the fluid to dissolve in water or other liquids
toxicity	poisonous content, which can cause sickness or death to animals or people – drilling fluids should have low toxicity

Unit 6

Problems while drilling	
shale sloughing	(pr. /sluffing/), a process by which shale, a common sedimentary rock, forms balls of material in the formation which are difficult to remove
blowout	an uncontrolled flow of fluid from a well
fish	a tool or broken pipe that has fallen to the bottom of the well
stuck pipe	a drillpipe stuck to the sides of the well
differential pipe sticking	stuck pipe caused by suction
dogleg	a sharp turn in a well
lost circulation	when mud escapes from the well
embrittlement	when metal becomes brittle because of a chemical reaction with fluid or gas
kick	when water, oil, gas, etc., enters the well because the pressure exerted by the mud is not sufficient to counter-balance the pressures in the well
corrosion	when metal is destroyed or damaged by a chemical reaction
Consequences of problems	
downtime	when no drilling can take place because of a problem
fatal	causing death
suspend drilling	stop drilling
money wasted	money lost unnecessarily
Preventing or solving problems	
act swiftly	act quickly and efficiently
retrieve the fish	bring the fish back to the surface of the well
latch onto a fish	attach the fishing string to the fish in order to retrieve it
jarring	trying to free a stuck pipe by creating a sudden movement in the string (a jar is sub-used for this purpose)
spotting fluid	fluid which is placed in a specific zone of the well to help free stuck pipe
reamer	a piece of equipment, used in fishing or for freeing stuck pipe, with a grinding tool which can widen the diameter of the hole
inhibit sloughing	prevent or reduce sloughing
spiral-grooved collars	a drill collar with spiral-shaped cuts to facilitate movement of the string through a narrow section of the well
brine	salty water used in the composition of some drilling fluids
spear	a tool used for fishing, which works by 'spearing' the fish and latching on from the inside
overshoot	a tool used for fishing, which works by latching onto the outside of the fish
kill a well	when there is a kick, use drilling fluid to control the pressure of the formation

Section 3
Drilling Operations

7 Preventing Blowouts

A blowout occurs when there is abnormally high pressure in the well, which leads to a kick as formation fluids enter the bore. The kick can usually be detected through changes in the level or flow of drilling mud. It is controlled by regulating the flow of mud and, if that fails, by closing valves on the blowout preventers (BOP). Once a blowout actually occurs, more radical action is required to cap the well. This unit focuses on vocabulary which will help you to describe what happens in the build-up to a blowout, and its consequences, in English.

Focused reading

1 Pressure rising

Read this case study of a blowout. In what order did these events take place?

1 A well control company was called in.
2 The rig and platform were evacuated.
3 The well caught fire.
4 The well was killed.
5 Weighted drilling mud was circulated.

Blowout on Ensco 51

Working offshore Louisiana, the crew of the Ensco 51 had just set casing when the well experienced an uncontrolled flow, causing a blowout and fire.

Attempts were made to stem the increasing flow of gas and fluids from 10+ valve using sea water and weighted drilling mud, but sufficient volume could not be added to slow the flow. The flow continued to increase and the decision was made to abandon both the rig and platform at 0300 hours. The 10+ valve was opened to divert the flow of gas away from the rig and platform, and all 43 personnel were then safely evacuated via two lifeboats to Platform B.

Specialist personnel from Wild Well Control were called in to help with capping and killing operations, but the gas flow ignited in the early hours of 02 March, due to an unknown ignition source. The well partially bridged on 03 March, causing the fire to go out, and the well was subsequently killed. Production on the platform had been shut in during drilling ops, which probably prevented a greater fire.

During the fire, the Ensco 51's derrick and substructure were completely destroyed, with the derrick collapsing onto the platform. The platform and its production equipment were also extensively damaged.

2 The fallout

José Miguel is a toolpusher on a rig in Alaska. His chief engineer, Paul Downey, is in Houston, Texas, at a meeting. Paul has just taken a break and gets a call from José Miguel. What are the differences between this blowout and the one described in the Ensco 51 case study? Which incident has the more serious consequences?

PD: José Miguel. How are you doing?
JM: Not too good, Paul. We've got a big problem.
PD: Go on.
JM: Well, we've got a blowout on Number 5 rig.
PD: Number 5? I thought we'd completed on 5.
JM: No, not quite, we were just tripping out when we got a big kick.
PD: Is anybody hurt?
JM: Well, there are no fatalities at least. But Okie's got a broken arm and both the others have got cuts and bruises.
PD: And you're OK?
JM: I'm fine, Paul, but it's a real mess here.
PD: And what about damage?
JM: Yes, there's quite a lot. One corner of the derrick's blown off and a big chunk of the drilling floor has gone.
PD: Any fire?
JM: Not so far, but we're keeping a close watch.
PD: So, what happened exactly?
JM: Well, the string was around 200 metres from the surface and we got a warning from the loggers. We circulated kill mud, but it was obviously too late.
PD: Did you manage to close the BOP?
JM: No, we were just about to close it when it blew.
PD: And what's happening at the moment?
JM: Well, there's mud everywhere. The fire crew are standing by and the medics are looking after the guys. I've called for the helicopter to take Okie off to hospital.
PD: Have you called well control?
JM: No, I wanted to call you first. I'll do that right away.
PD: José, I've got to get back to my meeting. I'd better tell them what's going on – they're paying after all! You're doing a great job. There's nothing more you can do. Keep me informed, won't you.
JM: Yes, of course. Bye.

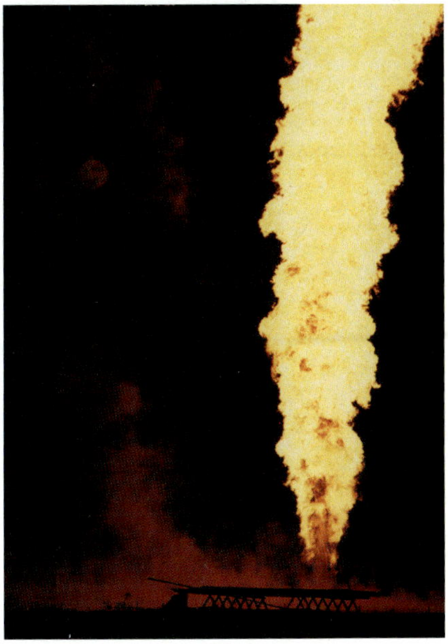

Blowout on a natural gas drill, California

Vocabulary-building activities

1 What happened?

These sentences are connected to the situations described in Focused reading 1 and 2. Choose the best verb to complete each sentence.

1 The platform was ________________. (bridged / killed / abandoned)
2 The well was ________________. (killed / circulated / ignited)
3 Weighted mud was ________________. (capped / circulated / blown off)
4 The crew were ________________. (abandoned / evacuated / damaged)
5 The derrick was ________________. (hurt / opened / destroyed)
6 The valve was ________________. (completed / cased / closed)
7 The roughneck's arm was ________________. (broken / damaged / destroyed)
8 The decision was ________________. (done / made / produced)
9 The gas was ________________. (ignited / sparked / blown)
10 The well control team was ________________. (called off / called in / called forward)

2 Before, during or after?

Look at the following sentences. In each sentence there are two actions. Does the second action happen before, during or after the first action?

1 The crew had just set casing when the well experienced an uncontrolled flow.
2 As the flow continued to increase, the decision to abandon the platform was taken.
3 Production on the platform had been shut in during drilling ops, which probably prevented a greater fire.
4 We were just tripping out when we got a big kick.
5 We circulated kill mud when we got a kick warning from the loggers.
6 We were thinking about closing the BOP valve when the well blew.

3 Blowout preventer

This is a description of a blowout preventer (BOP) from an oil technician's training manual. Complete the extract with the following types of ram:

blind seal rams   pipe seal rams   shear rams   slip rams

Blowout Preventer

A blowout preventer, usually known as a BOP, is a large, complex valve situated on the surface at the top of the well that can quickly be closed should the drilling crew lose control of formation fluids. The valve is operated by hydraulic actuators and once it is closed, the drilling crew can usually regain control of the reservoir and begin to increase mud density. As soon as density has been normalized, the BOP can be reopened and the formation can be returned to its optimum pressure.

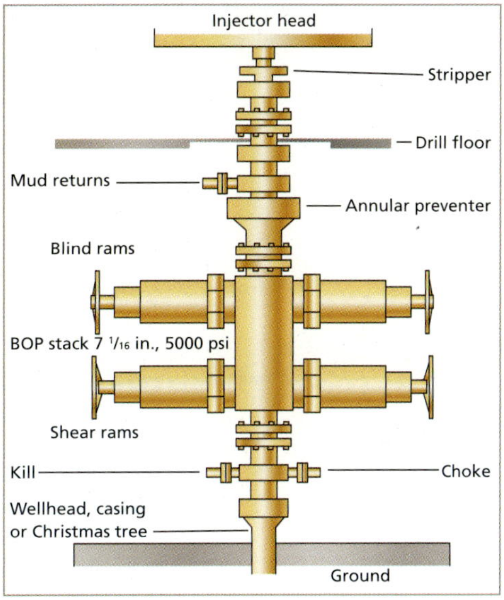

BOPs are found in a range of styles, sizes and pressure ratings. They can be configured with one or more of the following types of rams:

- A set of rams that seal around tubing, isolating wellbore below, known as **(a)** ____________
- A set of rams that seal off well pressure when there is no tubing in the wellbore, known as **(b)** ____________
- A set of rams to cut through tubing up to 3 ½" in diameter, known as **(c)** ____________
- A set of rams that hold the tubing in place during shearing or other applications, known as **(d)** ____________

BOPs are vital to drilling safety and so they are inspected and maintained regularly. Tests vary in frequency according to the likelihood of blowouts, from daily on wells where control issues are considered a high risk, to monthly or less frequently where well control problems are considered unlikely.

Refresh

Refresh the vocabulary you have learnt in this unit.

Put it to work

Have you experienced a blowout or do you know about a blowout which has occurred? What exactly happened? Describe the sequence of events in as much detail as possible.

Section 3
Drilling Operations

8 Optimized Drilling

Drilling technology has developed at an extraordinary pace in recent decades. This unit focuses on vocabulary which will help you to describe in English the new technologies that have optimized drilling efficiency.

Focused reading

1 New technologies

The following five extracts are from the website of the American Petroleum Institute (commonly referred to as the API), the main US trade association for the oil and natural gas industry, representing about 400 corporations involved in production, refinement, distribution and many other aspects of the industry.

1 Read each extract and underline the names given to the new technologies described.

Extract 1

With conventional drilling, oil and natural gas wells were drilled vertically at depths ranging from a few thousand feet to as deep as five miles. But new directional drilling (also known as slant drilling or deviation drilling) and horizontal drilling technologies allow drills to deviate from the vertical plane and go horizontal – or beyond. For oil and natural gas producers, this means reaching reservoirs that are not located directly beneath the drilling rig, and avoiding sensitive surface and subsurface environmental features.

Extract 2

Advances in directional drilling now permit multilateral drilling, where multiple offshoots of a single wellbore radiate in different directions and can contact resources at different depths. Development of this technology is recent and rapid, and promotes the use of one site instead of many sites.

Extract 3

At the same time, drilling away from the vertical plane has produced new challenges in extracting oil and natural gas from sophisticated drilling units. Coiled tubing technology has helped the industry meet this challenge. Coiled tubing is a continuous-length hollow steel cylinder of varying widths. Stored on a reel, coiled tubing is flexible, contains no joints and can be uncoiled or coiled repeatedly as needed. Since a significant percentage of oil production takes place from costly offshore platforms, coiled tubing has proven to be especially economical. The equipment can be set up quickly, it runs at higher speeds and it can work in existing wells.

Extract 4

But it isn't just the direction of the drill bit that has been revolutionized. Every time a drill bit enters rock below the surface, it displaces bits of rock called 'cuttings,' which then become rubble. Recent advancements in slimhole drilling have significantly reduced cuttings volumes. As the name suggests, the slimhole drill is smaller and displaces less rock. For example, a slimhole drilled to more than 2 ½ miles in depth and ending with a 4 ⅛-inch-diameter bottomhole, produces one-third fewer cuttings than a standard well at the same depth.

Extract 5

Both coiled tubing and slimhole drilling enable less disruptive, quieter drilling operations, minimizing the noise for wildlife or humans near the well site. Since coiled tubing is a continuous pipe, most noises associated with conventional drilling pipes are avoided. Efficient insulation and the equipment's smaller size further reduce noise levels. For example, the noise level of a conventional rig from a quarter-mile away is 55 decibels (less than the hum of an air conditioner), while a coiled tubing unit's noise level at the same distance is just 40 decibels (27% quieter). The smaller size of coiled tube drilling also cuts fuel use and reduces gas emissions when compared with traditional drilling.

2 Match one of these summaries to each extract.

This extract is about ...

- a benefits to the environment of new technologies.
- b drilling in wells with a very small diameter.
- c drilling formations which are not directly beneath the rig.
- d drilling with flexible piping instead of steel joints.
- e drilling several formations at the same time from the same rig.

Vocabulary-building activities

1

Technical diagrams

The technologies described in the Focused reading extracts are illustrated in the following photographs and diagrams. Label each photograph or diagram.

1 ____________________

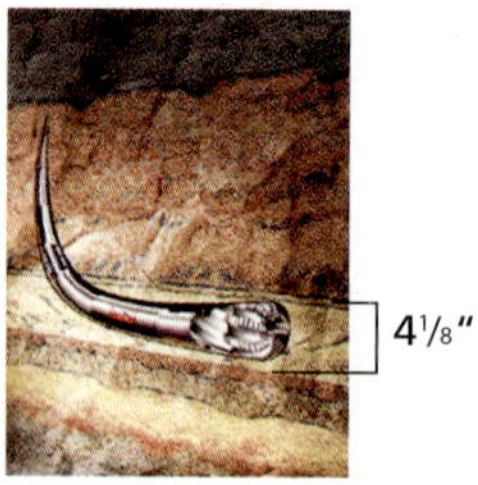

Source: www.slimhole.org

2 ____________________

3 ____________________

4 ____________________

5 ____________________

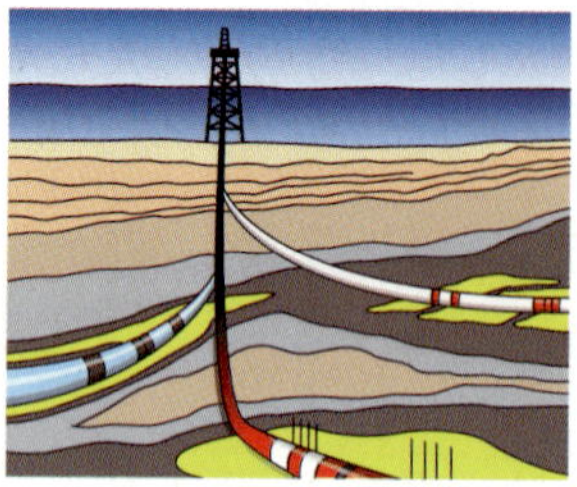

Source: RPS Group PLC (www.rpsgroup.com)

6 ____________________

2

Innovation and progress

Make comparisons between yesterday's technology and the more efficient technologies of today by completing the following sentences. Use the phrases in the box.

much better insulated   more effectively   more economical   more easily avoided
less disruptive   four times greater   fewer cuttings   far quieter   far less noise
easier to transport   better for the environment

1 Directional drilling is ____________________ because sensitive formations can be ____________________.

2 Multilateral drilling enables production rates up to ____________________ than conventional drilling.

3 Coiled tubing has proved to be ____________________ than conventional piping because it is ____________________ and reuse.

4 Coiled tubing also produces ____________________ because it is ____________________.

5 Slimhole drilling allows oil and gas companies to comply with environmental laws ____________ because it produces ____________.

6 Slimhole and coiled tubing drilling are ____________ and ____________ than traditional methods.

3 Joined-up thinking

These sentences describe a method of drilling called underbalanced drilling (UBD). Use a dictionary to check the meaning and use of the words in the middle column and then make a suitable connection. For 1–4, write two sentences for each. For 5–7, write one sentence for each.

		Connecting words	
1	In conventional drilling, formation fluids are prevented from entering the well as far as possible.	Therefore, However,	in UBD, formation fluids are allowed to flow into the well.
2	UBD is quite an expensive method of drilling.	Thus,	it is widely used because it reduces formation damage.
3	In UBD, the pressure in the well is kept lower than the pressure of the formation.	Nevertheless,	fluids are forced to flow into the well.
4	In UBD, an inert gas is injected into the drilling mud to reduce its density and hydrostatic force throughout the well depth.		the pressure of the well is reduced to below that of the formation.
5	In conventional drilling, formation fluids are prevented from entering the well as far as possible,	so that whereas	in UBD, formation fluids are allowed to flow into the well.
6	UBD is quite an expensive method of drilling,	yet	it is widely used because it reduces formation damage.
7	In UBD, the pressure in the well is kept lower than the pressure of the formation,		fluids are forced to flow into the well.

Refresh

Refresh the vocabulary you have learnt in this unit.

Put it to work

Have you had experience of any of the technologies described in this unit? Or of any other new technologies in drilling? Describe your experiences and the benefits that can be gained from these technologies, compared with conventional drilling techniques.

Glossary

Unit 7

Describing a blowout	
abnormally high pressure	pressure which is higher than normal
get a kick	the result of abnormally high pressure; when formation fluids flow into the well
uncontrolled flow of fluids	when the flow cannot be controlled
the well bridged	the well collapsed, creating a bridge
the gas flow ignited	the gas flow caught fire
the well blew	the blowout started, suddenly and dramatically
Preventing a blowout	
weighted mud	mud with additives to increase its density and pressure
stem the flow	to slow the flow of fluids
keep a close watch	watch carefully
production had been shut in	a valve had been placed in the well to control pressure
blowout preventer (BOP)	a series of valves at the top of the well to control pressures when there is a risk of a blowout
hydraulic actuators	devices which control the BOP hydraulically
seal/seal off the well	to close the well so that fluids cannot escape
a ram	in the blowout preventer, the rams are the components which quickly close to seal off the well (a ram is also a male sheep; the rig components come together like two rams fighting!)
blind seal ram	a ram used to close an empty well
pipe seal ram	a ram shaped to allow sealing around pipes or string equipment
shear ram	a ram made of hardened steel which will cut through piping and seal the well
slip ram	a ram designed to hold piping in place so that it can be sheared more efficiently
Regaining control after a blowout	
well control company	a service company called in when there is a blowout
capping	sealing off the well when there is a blowout
kill the well	to reduce the pressure in the well back to acceptable levels
kill mud	weighted mud used to reduce the pressure during a blowout
After a blowout	
the rig was evacuated	all the workers were taken from the rig
the platform was extensively damaged	the platform was widely damaged
cuts and bruises	superficial injuries caused by the blowout
a chunk of the drilling floor has gone	a large part of the drilling floor has gone

Unit 8

New drilling technologies	
conventional drilling	drilling in the traditional way, as it has always been done
directional drilling	drilling a well at an angle
slant drilling	another word for directional drilling (when something slants, it is at an angle between vertical and horizontal)
deviation drilling	another word for directional drilling (when something deviates from something else, it moves away from it; e.g., deviate from a straight line)
horizontal drilling	drilling along the pay zone in parallel to the formation bed at approximately 90 degrees to the well
multilateral drilling	drilling with several branches (or 'laterals') drilled out from the main well
coiled tubing drilling	drilling with a long, continuous, flexible length of tubing that comes wrapped around a reel
slimhole drilling	drilling with a small diameter wellbore (from 6¾ inches to 4¾ inches)
underbalanced drilling (UBD)	drilling where the pressure of the mud is lower than the pressure of the formation, allowing fluids from the formation to flow into the well
develop at an extraordinary pace	develop very quickly, at high speed
gain benefits from	to get advantages or profits from
deviate from the vertical plane	move away from a line directly down from the drilling rig
multiple offshoots	many branches coming from (or 'shooting out from') the main well
hollow steel cylinder	a pipe made of steel, with empty space through the core
rubble	pieces of broken rock
displaces less rock	breaks up and moves rock to another place
less disruptive operations	operations which have less negative impact (here on the environment)
the hum of an air conditioner	the continuous noise made by an air conditioning unit
decibels	a unit of measurement for noise
Connecting words: comparing and contrasting	
however nevertheless on the other hand	connecting words used to connect ideas in separate sentences
although yet whereas	connecting words used to connect ideas within the same sentence
Connecting words: giving causes, reasons, purposes, etc.	
therefore thus as a result	connecting words used to connect ideas in separate sentences
so that hence since because	connecting words used to connect ideas within the same sentence

Section 4 Contents

Geological and Geophysical Operations

	Titles	Focused Reading	Vocabulary Areas	Aim(s)
1	Formation Characteristics	• Transcript of a training presentation on basics of petroleum geology • Description of types of sedimentary rock	• Petroleum geology • Types of rock • Rock characteristics • Reservoir locations	To be able to describe the location and reservoir rock of a formation.
2	Petroleum Traps	• Transcript of a meeting between a petroleum geophysicist and his client	• Types of petroleum trap • Formation of traps	To be able to describe the different types and formation of petroleum traps.
3	Geological Consulting to the Oil and Gas Industry	• Covering e-mail sent by a petroleum geologist to a potential client • e-Sales brochure advertising the services of a petroleum geologist	• Commercial aspects of geological consultancy • Service description • Informal/formal communications	To be able to use commercial language to persuade a client to invest.
4	Petroleum Mapping	• Extracts from a presentation about maps, cross sections and the latest technology used for mapping • Maps and cross-sectional diagrams	• Features of maps and cross sections • Direct and indirect questions	To be able to describe the use and interpretation of different types of petroleum maps.
5	Seismic Surveying	• Article from a petroleum industry magazine	• Seismic acquisition • Sources and receivers • Principles of seismology	To be able to describe the use and interpretation of seismic surveys.
6	Mud Logging	• Mud log diagram • Website promoting mud logging company services	• Mud logs and mud logging • Well log terminology	To be able to describe the use and interpretation of mud log data.
7	Wireline Well Logging	• Introduction to wireline logging for trainee field engineers • Diagram of a sonde • Types of wireline well log	• Sonde anatomy • Uses of wireline well logs • Alternatives	To be able to describe the use and interpretation of wireline well log data.
8	Anatomy of an Oilfield	• Overview of a Saudi Arabian oilfield • Newspaper article about Siberian oilfield development	• Review of geological and geophysical vocabulary of previous units	To be able to produce a report or give a detailed presentation of an entire oil and gas field.

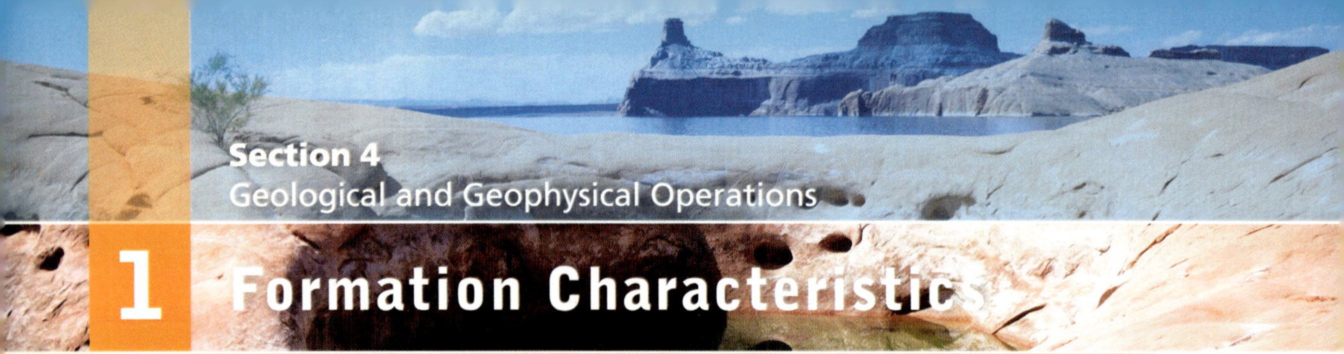

Section 4

Geological and Geophysical Operations

1 Formation Characteristics

Hydrocarbons – oil or gas – migrate through a rock formation from the source to the reservoir. Migration depends upon the type of rock, its characteristics and location. Knowing how much oil and gas is present and how easily it will flow on production are key elements in understanding whether or not a reservoir will be economic to exploit. This unit focuses on vocabulary which will help you to describe formation characteristics in English.

Focused reading

1 **Viability assessment**

Fedor is a petroleum geologist based in Siberia, whose job involves identifying new reserves and mapping existing reserves using geological data. In the following extract, he is explaining some of the basics of petroleum geology to a group of trainee field engineers.

Fedor mentions two factors: 1) porosity and permeability; and 2) water saturation. In relation to these factors, what is the difference between an economic and an uneconomic reservoir?

If we suspect that hydrocarbons might be present in an area, we commission an exploration well to obtain further information. These are also called wildcat wells. Using various tools, we are able to analyze the formations in great detail to establish whether or not it will be worthwhile producing.

First of all, we need to establish where the hydrocarbons may have come from – the source rock – and analyze the organic content, known as kerogen. Then, we look at the reservoir formation – that is, the rock into which the oil or gas has migrated and where it accumulates in a geological trap of some kind.

The most important aspects of a reservoir formation are its porosity and permeability. Most rocks, in particular sandstones and conglomerates, have at least some porosity. If enough pores are present, the pores are large enough and the pores are interconnected so that fluids flow through them (i.e., the rock is permeable), then the rock is a potential petroleum reservoir. With sandstones, a permeability of

Porosity: A measure of how much pore space is between the rock grains.

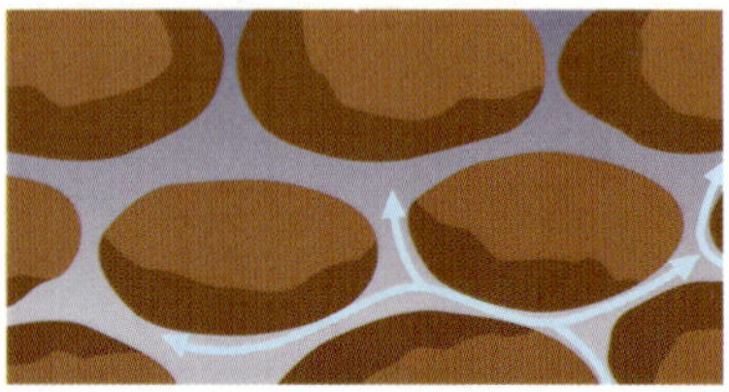

Permeability: A measure of how well the pore spaces are connected by pathways.

18% or more is usually needed for an economic oil reservoir. Gas flows easier than oil, so as little as 12% permeability may be enough for a gas reservoir. Porosity and permeability are important, but a petroleum reservoir needs to contain hydrocarbons as well. In water-bearing reservoirs, the pores are filled entirely with a salty solution called formation water, but in oil- and gas-bearing reservoirs, some oil or gas is present as well. A general rule of thumb is that 40% or more of the pore fluids must be hydrocarbons (i.e., the water saturation is less than 60%) in order for the reservoir to be economic enough to produce. If the water content is greater, then the oil tends to stay behind and the reservoir produces only water. These types of reservoirs are said to be 'wet'. If the water saturation is less, then the reservoir may be 'productive'. Whether or not it will be 'economic' to produce and make any money will depend not only on the rate of production, but also on how long the well will produce, in combination with many other factors, not least of which is market price.

2

Sedimentary rock

Read this description of the three main types of sedimentary rock and complete the table below.

Sandstone is composed primarily of sand grains that have been naturally cemented together. It is rough to the touch and can be white to buff to dark in colour. Sandstones are commonly deposited on beaches, river channels or dunes. Sandstone is a common reservoir rock for gas and oil and is the most important reservoir rock in North America.

Shale is composed of clay-sized particles and is the most common sedimentary rock. It is relatively soft and the colour ranges from green and grey to black, depending on the organic content. Darker shales are common source rocks for oil and gas, commonly deposited on river flood plains, and on the bottom of oceans, lakes or lagoons. Some shales are now being treated as reservoirs in North America.

Limestone is composed of calcite mineral grains that range in size from very fine to large sparkling crystals. The rock is commonly white or grey in colour. Limestone is a common reservoir rock and is typical in desert areas, such as the oil and gas fields of the Middle East. An organic-rich, dark-coloured limestone can also be a source rock for oil and gas.

Many sedimentary rocks are a combination of these three types, and mixtures are described as either sandy, shaly or limey (calcerous). For example, sandy shale is a shale with a high sandstone composition. Coal is also increasingly being exploited as a gas-bearing reservoir, having originally been a source rock, as in the southern North Sea.

	Source, reservoir rock or both?	**Typical colour(s)**	**Typical location(s)**
Sandstone			
Shale			
Limestone			

Vocabulary-building activities

1

Geological formations

In the box below are six aspects of a geological formation studied by a petroleum geologist. Fill in the blanks in the sentences. In two of the sentences, the word is used twice – once singular and once plural.

maturation   migration   reservoir   seal   source   trap

1 Evaluation of the ________________ uses the methods of geochemistry to quantify the nature of organic-rich rocks which contain the precursors to hydrocarbons, such that the type and quality of expelled hydrocarbon can be assessed.

2 The ________________ is a porous and permeable lithological unit or set of units that holds the hydrocarbon reserves. Analysis of ________________ at the simplest level requires an assessment of their porosity (to calculate the volume of in situ hydrocarbons) and their permeability (to calculate how easily hydrocarbons will flow out of them).

3 The ________________, or cap rock, is a unit with low permeability that impedes the escape of hydrocarbons from the reservoir rock. Common ________________ include evaporites, chalks and shales.

4 The ________________ is the stratigraphic or structural feature that ensures the juxtaposition of reservoir and seal, such that hydrocarbons remain in the subsurface rather than escaping (due to their natural buoyancy) and being lost.

5 Analysis of ________________ involves assessing the thermal history of the source rock in order to make predictions of the amount and timing of hydrocarbon generation and expulsion.

6 Finally, careful studies of ________________ reveal information on how hydrocarbons move from source to reservoir and help quantify the source (or kitchen) of hydrocarbons in a particular area.

2

Word formation

Complete this table using the different forms of each word. Most of them can be found in the Focused reading extracts on the previous two pages.

Verb	Adjective	Noun
		sediment
	porous	
permeate		
		migration
	saturated	
accumulate		
		fracturing
expel		

3 Geographical features

The photographs below show a number of geographical features where subsurface oil and gas reservoirs can often be found.

1 Which three geological features, from among those defined below, are shown?

a ____________

b ____________

c ____________

2 Read the following definitions again and decide whether operations in such environments would be onshore drilling or offshore drilling.

a A *basin* is a large area with thick sedimentary rocks, covered by shallow tropical water. In some cases over time, the basin has been filled by sediment and become dry land.
b *Reefs* are mounds of shells separated from the land by a body of water known as a *lagoon*. Some ancient reefs are now landlocked, created when the sea evaporated.
c A *coastal plain* is a strip of land adjacent to the ocean.
d *Shorelines* or beaches are long narrow deposits of well-sorted sand.
e *Sand dunes* are formed by wind in both desert and coastal environments.
f A *trench* is a deep channel in the seabed.
g A *ridge* is a submarine range of mountains, including volcanoes, which sometimes appear above the surface of the ocean.
h A *delta* is a mass of sediments deposited by a river flowing into a large body of water, such as a lake or ocean.
i A *continental shelf* is a shallow platform extending into the ocean from the beach.
j A *continental slope* is steeper than a shelf and lies between the shelf and the ocean bed.
k A *submarine canyon* is a narrow eroded channel through a shelf or slope.

Refresh

Using one of the strategies for consolidating and storing new vocabulary recommended in the Introduction, refresh what you have learnt in this unit.

Put it to work

Choose a geological formation that supports an oil reservoir your company is exploiting. Describe the conditions that made the reservoir viable for exploitation. Where is the reservoir located? What is the source rock formation? What type of rock is the reservoir rock? What is its porosity and permeability or water saturation?

Section 4
Geological and Geophysical Operations

2 Petroleum Traps

Hydrocarbons migrate from their source through any rock formations that are porous and permeable. If this migration is halted by a geological fault or an impermeable rock seal, a trap is formed. The hydrocarbons can accumulate in the trap, creating a reservoir. This unit focuses on vocabulary which will help you to describe the formation and characteristics of petroleum traps in English.

Focused reading

1 Decision time

The following transcript is the record of a meeting in which a geophysicist is discussing a petroleum trap with a manager from an oil company, with reference to the diagram, opposite. They are deciding whether to buy a lease to develop a potential oilfield at Carson Sound in Alaska.

1. Why does the geophysicist describe the geology of Prudhoe Bay?
2. How is the field sealed on each side – north, south, west and east?

Manager: So, can you describe the geological formation of Carson Sound?

Geophysicist: Sure. As we have explained in our report, this field appears to be an extension of the Prudhoe Bay field. If I could describe the geology of Prudhoe Bay first of all, it will help me to explain the geology of Carson Sound. This diagram shows the central North Slope. As you can see, Prudhoe Bay is the result of a combination of structural and stratigraphic circumstances. They were formed by unconformities created when the rocks were uplifted during rifting and then buried again.

Manager: So, how is the accumulation trapped?

Geophysicist: Well, on the north side of the bay the field is sealed by a normal fault, and in the south by south-dipping shales, which overlie the reservoir sand. The western extent of the oil is limited by a structural saddle in the upper surface of the reservoir. Unconformities and overlaying shales seal the crest and the eastern flank of the oilfield.

Manager: So Carson Sound is part of the same formation, is it?

Geophysicist: Well, the oil is certainly from the same source, but there are one or two crucial differences.

Manager: Such as … ?

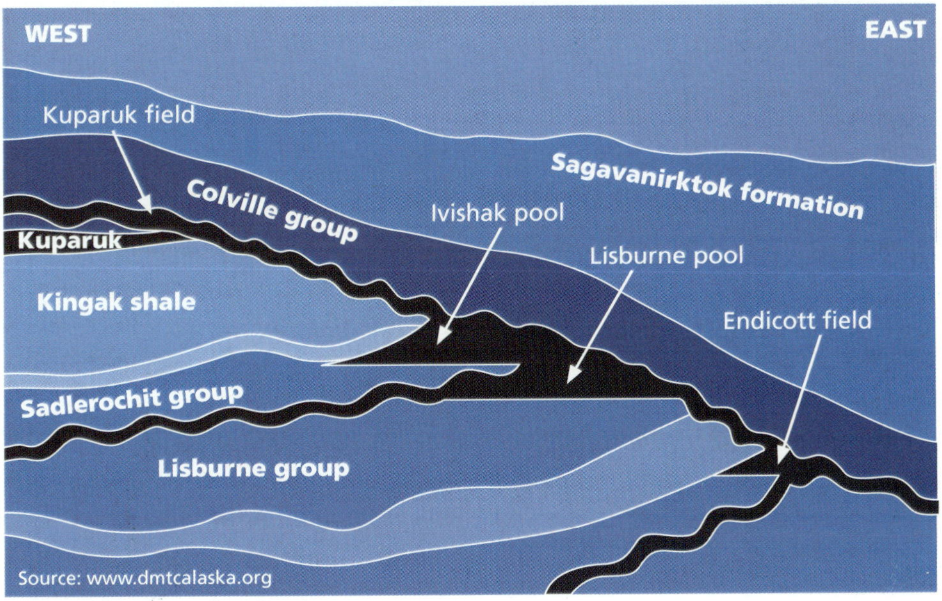

Geological formation of Prudhoe Bay

2

Risk assessment

Now read the continuation of the meeting.

1 What is the connection between Mukluk and Carson Sound?
2 Do you think that the manager should be worried?

Geophysicist: I have to tell you that there is always the risk that Carson Sound could turn out to be a Mukluk.

Manager: What do you mean, a Mukluk?

Geophysicist: Yes, I suppose that was before your time! Mukluk was one of the biggest financial disasters in oil and gas history – in 1983, nearly twenty years after the huge success story of Prudhoe Bay.

Manager: Go on.

Geophysicist: Well, Mukluk is just on the other side of Prudhoe Bay, and it was also an angular unconformity trap, like Carson Sound. Seismic data indicated a potential reservoir rock in the same Sadlerochit Sandstone with a shale cap rock. The lease was sold for an estimated $1.5 billion and 12 companies spent $100 million on building an artificial island in the Beaufort Sea, and then a further $150 million on exploration drilling. They were looking at anything between two billion and 10 billion barrels of oil, depending on who you believe.

Manager: Wow! So what happened?

Geophysicist: Absolutely nothing. It turned out that the oil had leaked out of the formation centuries ago. There was nothing left. It is still the most expensive dry hole ever!

Manager: Well, this isn't nearly as big, is it?

Geophysicist: No, but you still don't want to waste your money, do you?

Manager: And what are the chances of that happening?

Vocabulary-building activities

1

The geology of petroleum traps

Read this description of two different types of petroleum traps and label the diagrams. The main features are given in the box below.

anticline   fault trap   pinch-out   salt dome   truncation

There are two basic types of petroleum traps – structural and stratigraphic.
Structural traps occur when reservoir rock layers are deformed, such as in a fault or a fold. In a fault trap, a rift occurs in the rock, causing the layers to slip and tilt in one direction or another at the point of deformation, meaning that the migration of oil through a particular layer is halted. There are two types of folds – synclines and anticlines. Of these, normally only anticlines become petroleum traps. An anticline occurs when rock layers form an arch-like structure and the petroleum migrates to the highest point and is prevented from moving further by an overlying bed of impenetrable rock.

Stratigraphic traps can be formed by the deposition of reservoir rock, by an angular unconformity, or by a change in the porosity of the sedimentary layer. Examples of depositions include a river channel sandstone or a limestone reef encased in shale. One example of an angular unconformity is when an angular sedimentary layer is truncated, or cut off, by a horizontal layer. This is known as a truncation. Another, which is called a pinch-out, is where the sedimentary layer is trapped between two other layers when they meet at a point.

Petroleum reservoirs can also be created by traps which are both structural and stratigraphic. These are called combination traps. An example of this is a salt dome, where a layer of salt pushes upwards and deforms the surrounding layers.

1 salt dome

2 ____________

3 ____________

4 ____________

5 ____________

2 Angular unconformity traps

The two largest fields in North America, the East Texas field and the Prudhoe Bay field, are both in angular unconformity traps. The following series of diagrams and the accompanying descriptions of each stage of the process below explain how a petroleum trap is formed. Put the first five sentences in the correct chronological order. The final stage in the last sentence is in the correct place.

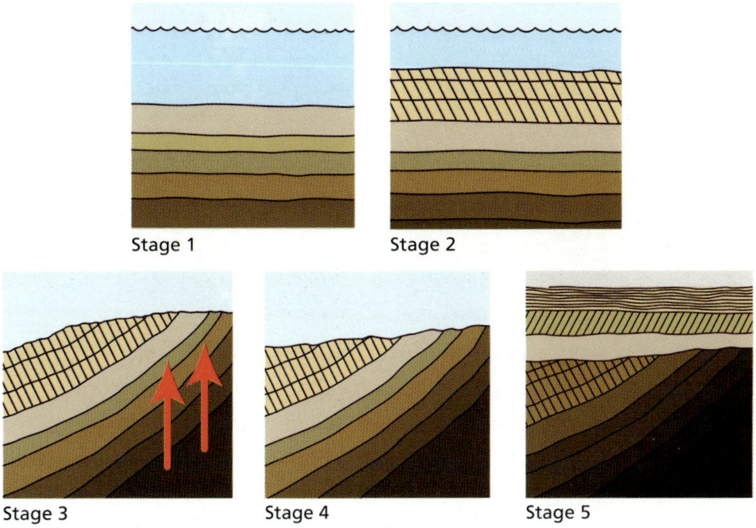

1 The seas again invaded the area and a layer of chalk was deposited on the angular unconformity. This would become the cap rock of the East Texas field.

2 The sandstone was buried beneath the subsurface as it became covered by a layer of sediments from the sea.

3 Later, the Sabine Uplift, a geological event along the Texas-Louisiana border, arched up and exposed the Woodbine sandstone above the surface of the sea.

4 Approximately 100 million years ago, East Texas was covered by shallow ocean. The Woodbine sandstone was deposited during this geological era. This would become the formation rock of the East Texas field.

5 The top of the arch of the Woodbine sandstone was eroded and an angular unconformity was created.

Eventually, petroleum migrated from its source in the Eagle Ford shale along the Woodbine sandstone and became trapped by the layer of Austin chalk, accumulating into a reservoir of some five billion barrels of oil.

Put it to work

Consider an oilfield that your company is working on or has worked on in the past. Describe how its reservoir was formed. What kind of petroleum trap was created? What made geologists confident that the field could be exploited?

Refresh

Refresh the vocabulary you have learnt in this unit.

Glossary

Unit 1

Petroleum geology	
source rock	the rock from which oil and gas is generated
reservoir rock	the sedimentary rock through which oil and gas is forced by the earth's pressure
porosity	the volume of the rock that is made up of gaps (pores), which can hold hydrocarbons (i.e., the volume not occupied by solid matter)
permeability	a measure of the ease with which fluids can travel through a rock formation
seal (or 'cap rock')	an impermeable rock layer at the top of an oil or gas reservoir
trap	an area of a formation where oil and gas can collect
accumulation	a gradual build-up of fluid in a geological formation
migration	movement of fluid under pressure from the source rock through the reservoir rock
maturation	the time taken for oil and gas to form and migrate (by which the age of a formation can be calculated)
saturated	describes a rock formation with a very high proportion of water
deposited	when geological material is put in a particular place by a force of nature, such as wind, river, sea, etc.
buoyancy	the fluid's ability to move naturally to the surface
Types of rock	
sedimentary rocks	rocks composed of particles or salts (sediment) deposited on the surface of the ground or on the bottom of the ocean
sandstone	a common sedimentary rock, mainly consisting of sand grains
limestone	a common sedimentary rock, composed of calcium, carbon and oxygen, in the formula $CaCO_3$
shale	a common sedimentary rock, composed of fine particles
clay	a common sedimentary material composed of different minerals and organic substances
chalk	a very fine-grained limestone (= calcium carbonate)
Geological formations	
basin	a natural depression where sedimentary rocks have accumulated
reef	a ridge just above or below the surface of the sea, made of coral, sand, etc.
lagoon	a stretch of sea water separated from the sea by a sandbank or reef
plain	a large area of flat land with few trees
shoreline	the line along which a large body of water meets the land
dune	a mound or ridge of sand formed by the wind
trench	a long narrow depression in the ocean bed
ridge	a long narrow mountain range, whether on land or on the ocean floor
delta	a triangular area of sediment deposited at the mouth of a river
continental shelf	a flat or slightly sloping area of land extending from the land under the ocean
continental slope	an area of land falling steeply to the ocean bed from the continental shelf
canyon	a deep gorge, on the ocean bed or on land, often with a river flowing through it

Unit 2

Petroleum traps	
trap	an area of reservoir rock contained by a cap rock, which prevents petroleum from migrating
halted	stopped/prevented from moving
structural trap	a trap formed by the deformation of rock layers
stratigraphic trap	a trap formed by deposition or erosion
unconformity	a rock surface created by erosion
deform	change the shape of a rock layer (deformation, _n_)
fault	an extended break in a rock formation, where there has been movement on either side
fold	a bend in a layer of sedimentary rocks
normal fault	a fault where the movement has been mainly vertical
overlie	literally 'lie over' – where one layer overlies another in a rock formation
dip	to move in a downwards direction
slip	to suddenly move downwards
tilt	to move upwards or downwards into a sloping position
south-dipping shales	a layer of shale at an angle extending towards the south
saddle	a curved layer of rock acting as a seal
angular unconformity	a type of stratigraphic trap where the layers have tilted at an angle
anticline	a type of structural trap, where the layers of rock are folded upwards (as opposed to a syncline, where the rocks are folded downwards, but cannot form a trap)
pinch-out	a type of trap where the sedimentary layer thins out and the layer above and below comes together (thins to zero) like a wedge
truncate / truncation	a type of trap where an angular (sometimes vertical) sedimentary layer is sealed by a horizontal layer of impermeable rock
salt dome	a large mass of salt which has flowed upwards through the sedimentary rocks, causing deformations in the surrounding layers
impenetrable	cannot be passed through (or penetrated)
encased	describes a reef that is surrounded on all sides by shale
Other geological terms	
the earth's crust	the rocky, relatively low density, outer layer of the earth
rift / rifting	a large horizontal fault in the earth's crust
buried	when one layer of rock is covered by another or by deposits
crest	the top of a hill, fold or dome
flank	the side of a formation or area of land, for example
deposition	another word for deposit
erosion	gradual wearing away of soil or rock by weather, sea, pollution, etc.
invade	describes when the sea comes to flood or cover the land

Section 4
Geological and Geophysical Operations

3 Geological Consulting to the Oil and Gas Industry

Oil and gas companies depend on petroleum geologists to advise them about the best places to look for hydrocarbons. Many petroleum geologists work as independent consultants for different oil and gas companies. This unit focuses on English vocabulary which will help you with the commercial aspects of the work of an independent consultant to the oil and gas industry.

Focused reading

1 First proposal

This is an e-mail written by a consultant petroleum geologist, Paul Selby, to the manager of the Exploration Division of a large oil and gas company in Qatar, Adel Al-Jalawi. What is the purpose of the e-mail?

Send Chat Attach Address Fonts Colors Save As Draft Photo Browser Show Stationery

To: Adel Al Jalawi (adel.al-jawali@qatarpetroleum.com)
From: Paul Selby (paul@selbygeoprospects.com)
Subject: O&G opportunities offshore West Africa

Dear Mr Al-Jalawi

I contacted you just over two years ago and you were good enough to allow me time to show you some data we had assembled on prospects in the Ngosso and Iroko license areas off the coast of Cameroon. In the end, you decided not to invest in this area, but as you will know, it has become apparent since then that there are some excellent prospects here. A number of other companies have drilled wildcats and over a dozen wells are now producing. We have continued to do research offshore of Cameroon, as well as of Gabon and Equitorial Guinea, and we have now teamed up with a seismic company, Exploration Geosciences. If you would be interested in seeing our findings on more than 20 prospects in offshore West Africa, I would be delighted to make an appointment to present and discuss them in greater depth. I have attached an updated list of our main petroleum geology services. I look forward to hearing from you.

Yours sincerely,
Paul Selby
Managing Director, Selby Geo-Prospects Ltd

2 Sales brochure

This is an extract from the attachment sent by Paul Selby. Which topic is relevant if a company would like ...

1 to have more information about areas which are already being drilled?
2 to know how much oil and gas is present in a possible oil and gas field?
3 to have detailed geological information about an oil and gas field?

These are the main services offered by Selby Geo-Prospects. Please contact us if there are any geological or geophysical services you would like us to supply.

Selby
Geo-Prospects Ltd

Regional evaluations

Countries, regions, basins, concessions. Analysis of tectonic setting, structural evolution, stratigraphy, geochemistry, petroleum systems, exploration history, reserves and future potential.

Open-acreage assessment

Exploration history, drilling results, discoveries, dry holes, reservoirs, seals, traps, source rocks, timing of migration and remaining potential.

Identification of farm-in opportunities

Evaluation of currently licensed blocks to assess remaining potential in terms of drilling results to date, unexplored petroleum systems, undeveloped reserves and yet-to-find potential.

Risk assessment and chance of success

Assessment of risk and chance factor based on reservoir, seal, trap and source rock parameters. Ranking according to degree of risk.

Prospect evaluations

Evaluation of reservoir, seal, trap and source rock characteristics, burial history and stochastic calculation of potential reserves.

Reserves assessment

For countries, regions, basins and concessions. History of discoveries, cumulative reserves, production, remaining reserves, reserves replacement, yet-to-find oil and ultimate endowment.

Studies of individual fields

Structure, size, reservoir characteristics, seal, source rock, reserves, drilling history, production history, reservoir behaviour, depletion and remaining reserves.

Technical reports

On the above topics or on a combination of the above topics.

Vocabulary-building activities

1

Selling the services

These are extracts from e-mails written by Paul Selby, the consultant petroleum geologist, to clients and potential clients. Which of the services listed in the sales brochure in Focused reading 2, on the previous page, is he writing about?

1. We have carried out research into the drilling and production which have taken place at this field and our findings are as follows: …
2. According to our calculations, the chances of success in each of these reserves are as follows: …
3. As you know, we have been operating in this region for some time and we have gathered a huge quantity of data that you might be interested in.
4. Approximately 70% of the block is currently licensed, but we believe that there are still exciting opportunities available among the remaining acreages.
5. We are pleased to accept your offer of contract to carry out evaluation of remaining potential in your F2 reserve offshore Equatorial Guinea.

2

Conditional reply

This is the reply that Paul Selby received from Adel Al-Jalawi. Fill in the gaps in the e-mail using the following words. Sometimes there may be more than one correct answer.

would should could may

To: Paul Selby (paul@selbygeoprospects.com)
From: Adel Al-Jalawi (adel.al-jalawi@qatarpetroleum.com)
Subject: RE: O&G opportunities offshore West Africa

Dear Mr Selby

Thank you very much for your e-mail. I remember our meeting and we were all impressed with the quality of your presentation and the detail of your research data. As I explained after we met, our reasons for deciding against investment in West Africa were political. At the time, we felt that the region lacked the stability we required to make a firm long-term commitment. I believe it was the correct decision and that it **(1)** ________ have been a very risky venture two years ago.

Of course, we have been monitoring the situation, but we remain cautious as to whether we **(2)** ________ invest there or not. Nevertheless, we **(3)** ________ be very pleased to hear your views and you **(4)** ________ be able to convince us to go ahead. We **(5)** ________ be available for a meeting sometime in the week beginning 16th Feb. Please **(6)** ________ you let me know if you **(7)** ________ be able to travel to Qatar at that time? Jamila, my secretary **(8)** ________ be pleased to help you make any arrangements you **(9)** ________ need.

Looking forward to hearing from you.

Yours sincerely,

Adel Al-Jalawi

3 Formal and informal style

Here are some sentences that say the same thing, but in a different style, according to whether they come from formal or informal e-mails. Complete the sentences with one suitable word in each space.

	Informal	Formal
1	**(a)** ________ a lot for taking the time to **(b)** ________ in the other day.	Thank you very much for taking the time to meet me on Wednesday.
2	**(c)** ________ us know what you think.	We would be very **(d)** ________ to hear your views on the subject.
3	Can you get **(e)** ________ to me a.s.a.p.	I would be very **(f)** ________ if you could reply as soon as possible.
4	Have a good trip!	May I **(g)** ________ you a successful trip!
5	**(h)** ________ ask if you need anything.	Please don't **(i)** ________ to ask if you require any **(j)** ________ assistance.
6	**(k)** ________ forward to seeing you soon.	We look forward to seeing you at your **(l)** ________ convenience.

4 Openings and closings

This is a selection of openings to e-mails.

Hi James, ... Dear Ms O'Hea, ... Dear Sirs, ...
Dear Olga, ... Hello Pete, ...

1 Which of these are informal and which of them are formal?
2 Which of the following closings could follow each opening?

Best wishes, ... Take care, ... Yours faithfully, ...
Kind regards, ... Yours sincerely, ...

Refresh

Refresh the vocabulary you have learnt in this unit.

Put it to work

Write a formal e-mail to a client of your company. Describe briefly the services your company offers and ask for a meeting to present some more information.

Section 4
Geological and Geophysical Operations

4 Petroleum Mapping

Maps and cross-sectional diagrams are used by petroleum geologists and geophysicists to illustrate the location of oil and gas reservoirs. Making and interpreting maps and cross sections is an essential part of their work. This unit focuses on vocabulary which will help you to describe the use and interpretation of maps and cross sections, or to discuss them with consultants in English.

Focused reading

1

Maps and cross sections

A group of junior field engineers is attending a presentation about the use of maps and cross sections in petroleum geology. Their trainer, Josef, explains the difference between maps and cross sections. Imagine you are one of his students. Read his explanations and take notes about maps, cross sections, reservoir models and mapping technology.

First, he gives a brief overview.

'The easiest way to explain the difference is to say that a map is a view taken from above the Earth's surface and you are looking down at it, whereas a cross section is a view taken by cutting vertically through the Earth's surface, as if you are looking at it face-to-face. Does that make sense?'

Then, he reads to his students a definition of a geologic map.

'A geologic map or geological map is a special-purpose map made to show geological features. Rock units or geologic strata are shown by colour or symbols to indicate where they are exposed at the surface. Bedding planes and structural features, such as faults, folds, foliations and lineations, are displayed with strike and dip or trend and plunge symbols, which give these features three-dimensional orientations.'

Source: www.en.wikipedia.org/wiki/Geologic_map

Then, he explains the difference between a map and a cross section and shows them an example of a cross section.

'A cross section shows a vertical section through a volume, whereas a map projects a view taken of the surface of the geological area – a bird's eye view. Cross sections are often used to show the different types and orientations of formations found under the surface. Here is an example of a very large cross section taken over a horizontal distance of several hundred miles.'

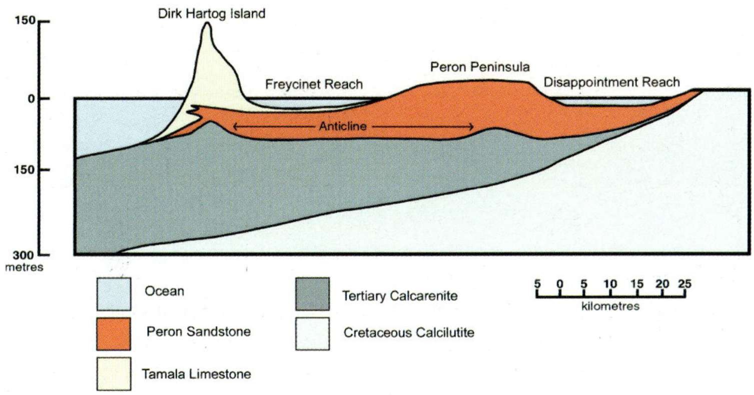

Source: © www.sharkbay.org

He goes on to talk about petroleum reservoir maps.

'To show a petroleum reservoir correctly, you need to have a number of different maps and cross sections to distinguish the various layers in the reservoir. The maps will show the structure of the formation, but also other important factors, such as gross and net thickness, porosity, water saturation and so on. A full set of maps and cross sections like this is normally called a reservoir description or a reservoir model.'

Finally, he talks about some of the latest technology used for the mapping of geological formations.

'Nowadays, petroleum geologists are increasingly using information and communications technology to manage and deliver geoscientific information, and this is driving the development of digital methods for geological data visualization, recording and interpretation in both 2D and 3D. An example of this is the British Geological Society's SIGMA project (System for Integrated Geospatial Mapping), which is updating geological survey methods to harness these rapidly developing innovations.'

Vocabulary-building activities

1

Interpreting different maps

Here are diagrams and descriptions of three different types of map. Read the descriptions and then follow these instructions:

1. Label each map correctly – topographic map, geologic map, isopach map.
2. On the topographic map, identify and label: **a)** contour lines; **b)** a steep slope; **c)** a gentle slope; d) the scale of the map.
3. On the geologic map, identify and label: **a)** strike; **b)** dip; **c)** angle of the plane.
4. On the isopach map, identify and label: **a)** sea level; **b)** pay zone.

This is an example of a topographic map, showing the elevations of an area where oil and gas may be present. The lines are contour lines showing the land elevation above sea level. This map shows a ridge with a gentle slope on the north-east flank and a steep slope on the south-west flank. You will notice that the greater the contour intervals, the gentler the slope. The scale on this map is 1 centimetre to 500 metres.

This is an example of a geologic map. It shows the orientation of the rock layers by defining strike and dip. Strike is the horizontal orientation of the formation, dip is the downward angle perpendicular to the strike, including the angle of the plane.

This is a type of subsurface map called an isopach map, which shows the thickness of the rock layers. For example, when an oil or gas field has been drilled, an isopach map can be used to describe the pay zone. This could be a gross pay map, which contours the thickness of the whole reservoir, including non-productive water-bearing or tight zones. Or it could be a net pay map, showing only the productive thickness.

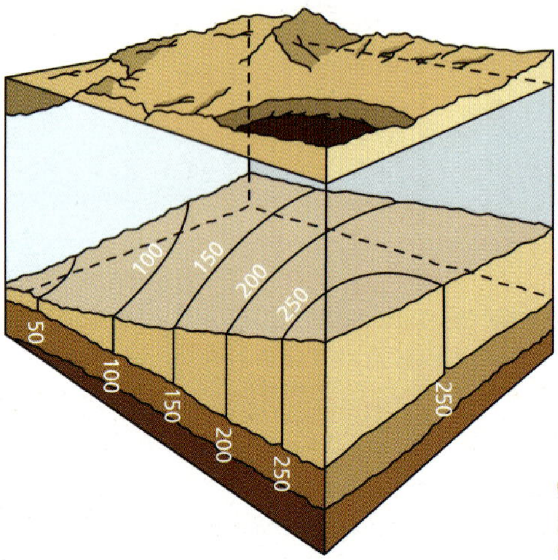

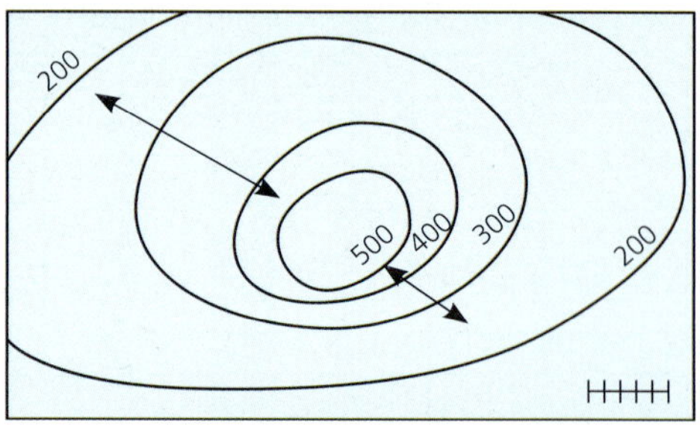

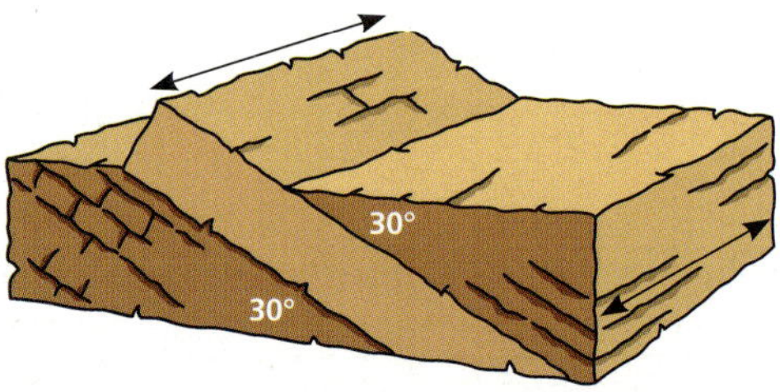

2 Any questions?

Following his presentation, the junior field engineers ask Josef a number of questions. Sometimes they ask the questions in the form of direct questions, such as 'How is a cross section made?' Sometimes they use indirect questions, such as 'Can you explain how a cross section is made?'

1 Complete the questions in the columns below.

	Direct questions	Indirect questions
1	What does a reservoir model consist of?	Can you tell us …
2		Can you show us how seismic data can be incorporated into cross sections?
3	What does SIGMA stand for?	Do you know …
4	How are old maps integrated with new ones?	Can you explain …
5		Can you show us how they capture field data?
6	What do you mean by GIS software?	Sorry, but I don't quite understand …
7	What is an isopach map?	Can you explain …
8		Could you tell me what a net pay map shows?

2 How many of these questions can you answer from the information on the previous page, or from your own knowledge?

Refresh

Refresh the vocabulary you have learnt in this unit.

Put it to work

Choose one of the types of petroleum map discussed in this unit and find an example online. What is the map's purpose? What information does it show? How was the map made?

Glossary

Unit 3

The commercial side of geological consultancy	
consultant	an expert in a specific area of work, paid by a company to act as adviser or participant in short-term projects on a freelance basis
prospects	potential commercial sources of oil and gas
teamed up with	joined with others for a particular purpose, job or project (also 'joined forces with')
concessions	possession of such gives the right to look for and/or produce oil and gas in a particular area
tectonic setting	the geological structure of the Earth's crust and its potential for producing resources such as oil and gas
open acreage	blocks of land which have not yet been licensed for oil and gas exploration
dry holes	wildcat wells which are unsuccessful (i.e., where there is no oil or gas)
remaining potential	after a certain period of production, the remaining potential is a calculation of the oil and gas which might be left in the reserve
farm-in opportunities	opportunities to get licenses to look for oil and gas
licensed blocks	blocks of land which have already been licensed (or 'farmed out') for oil and gas exploration
yet-to-find potential	see 'remaining potential'
risk assessment	calculation of the risk attached to an investment in oil and gas exploration
chance of success	measured as a percentage, this is a calculation of the likelihood that oil or gas will be found and can be developed
burial history	looking at the layers in a geological formation, one can study how the sedimentary rocks have been buried over time
stochastic calculation	a calculation based on statistics and probability, not on exact figures because of the random nature of the subject being studied
cumulative reserves	a measurement of the quantity of oil taken from a reserve to date
ultimate endowment	a measurement of the total quantity of oil taken from a reserve during its whole lifetime
depletion	occurs as petroleum is taken from a reserve and the remaining quantity of oil or gas decreases
risky venture	an investment which has a high probability of losing money
cautious	a safe(r) investment opposite of 'risky'
Informal and formal communications	
examples of informal phrases	Thanks a lot for … Let me know … Get back to me … Do ask if … Looking forward to seeing you.
examples of more formal phrases	Thank you very much ... We would be very pleased to … I would be very grateful if … May I wish you … Please don't hesitate to … I look forward to meeting you.

Unit 4

Maps and cross sections	
to map / mapping	to gather geological or geographical data and then use the data to create a map
cross section	a vertical view through a solid object (in this case, geological layers)
lineations	the lines between geological layers
foliations	where geological formations have split into thin sheets
trend	the horizontal direction of a geological feature
plunge	the vertical direction of a geological feature
scale	the ratio between the size of the map and the actual area the map shows; for example, 1 cm:250 m means that 1 centimetre on the map shows 250 metres actual size
signs and symbols	images used to represent physical features on a map
means of orientation	used to show directions on a map (north, south, etc.)
gross thickness	the thickness of a layer, including all its fluid contents: oil, water, etc.
net thickness	the thickness of a layer, including only its oil and gas content
topographic map	a map that uses contours to show the elevation of ground surface
geologic map	a map showing where rock layers appear on the surface of the Earth
isopach map	a map that uses contours to show the thickness of a rock layer below the surface
contour lines	lines on a map with the same value
gentle slope / steep slope	shown on a map by contour lines which are a good distance apart (as opposed to a steep slope, shown by contour lines which are very close together)
strike	the horizontal direction of rock layer or fault
dip	the angle and direction of a rock layer or fault, measured at right angles to the strike
pay zone	the area of a formation where oil and gas can be found
capture field data	to obtain data during a survey, often using digital technology

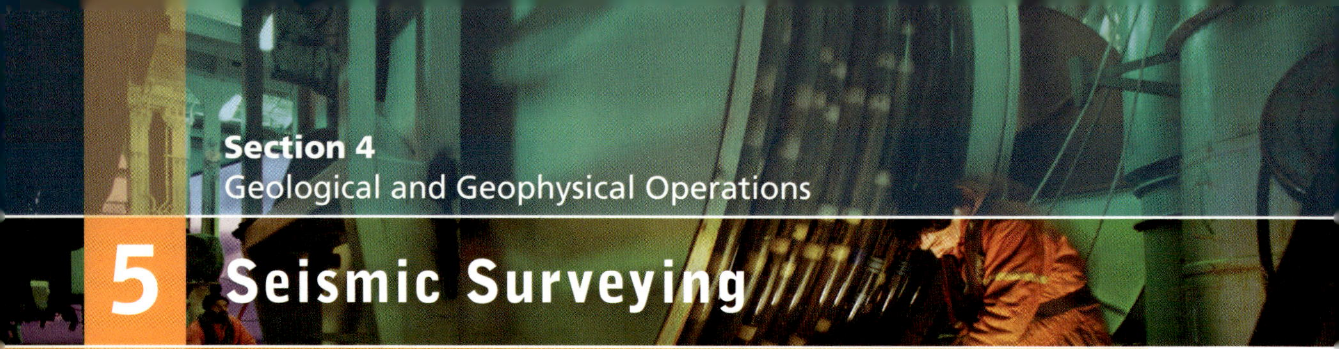

Section 4
Geological and Geophysical Operations

5 Seismic Surveying

Drilling a wildcat exploration well has always been a risky proposition, but advances in seismic surveying and data processing have refined the identification of what geophysicists call 'direct hydrocarbon indicators' to a point where the drilling risk has been substantially reduced. This unit focuses on vocabulary which will help you to describe the use and interpretation of seismic surveys in English.

Focused reading

1 Seismic exploration

Here is an extract from an article in a magazine called *Petroleum News* (www.petroleumnews.com) about developments in seismic exploration, in which the writer describes how seismic acquisition works.

1 How is natural gas detected in a seismic survey?
2 Why is it more difficult to detect oil?

Bright spots

In a seismic survey, sound waves from a sound source partially reflect off boundaries between different underground strata to form echoes that are detected at the surface by receivers called geophones. Those echoes provide information about the subsurface geology, including the locations of potential oil and gas traps.

Oil and gas occupying rock pores in an oilfield reservoir affect the physical properties of the rock in a way that could alter those sound echoes and thus provide direct evidence of subsurface oil and gas pools. For example, a quite modest amount of natural gas inside a rock will significantly reduce the velocity of sound passing through the rock. That velocity reduction can increase the acoustic contrast between the gas-bearing rock and the adjacent rock formations. And the increased contrast can, in turn, cause an abnormally high amplitude seismic reflection, giving rise to what geophysicists refer to as a 'bright spot' in a seismic section.

However, there can be more than one possible explanation for a seismic phenomenon such as a bright spot, and so this type of indicator suggests, but does not prove, the existence of subsurface hydrocarbons. And this is especially true when trying to use seismic data to detect oil. Oil has a much lower acoustic contrast with rock than gas. And, to make things even more tricky, there's quite a low acoustic contrast between oil and water, thus making these two liquids difficult to distinguish.

2

Developments in seismic technology

Here are two other extracts from the same article. Read the extracts and answer the questions.

1 How does AVO improve the performance of a survey?
2 What is the difference between 3D and 4D seismic?

AVO

Developments in seismic technology have enabled geophysicists to further reduce drilling risk. Here are two examples:

The first technique is known as amplitude variation with offset. AVO is a bi-product of the way in which a seismic survey involves recording underground sound reflections using sound sources and geophone sound detectors in a series of increasing offsets from a single survey point (an offset is the distance between the seismic source and the receiver). The seismic crew record the data from different offsets so that they can add the data together. This addition tends to remove random noise while enhancing coherent signals from underground sound reflections. AVO analysis is more successful in young, poorly consolidated rocks, such as those in the Gulf of Mexico, than in older, well-cemented sediments, such as those from the mid-continent region of the USA.

4D seismic

Another technique is known as 4D seismic. This technique involves shooting several 3D seismic surveys over the same area over a time period of perhaps several years (a 3D survey is a type of survey that results in a three-dimensional image of the subsurface geology). Changes in seismic signals from one survey to the next can provide insights into the movement of fluids, such as oil and gas within the field reservoir.

Vocabulary-building activities

1

Sources and receivers

In seismic acquisition, a source of sound energy and a receiver are required. Read the description of the use of a vibrator truck in seismic surveying and decide whether each sentence refers to the source (S) or the receiver (R).

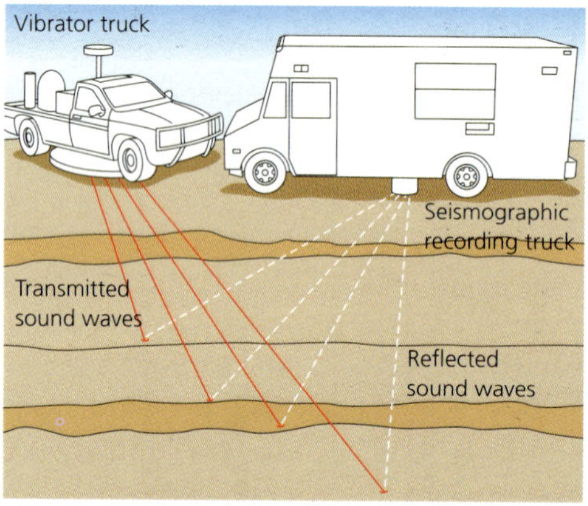

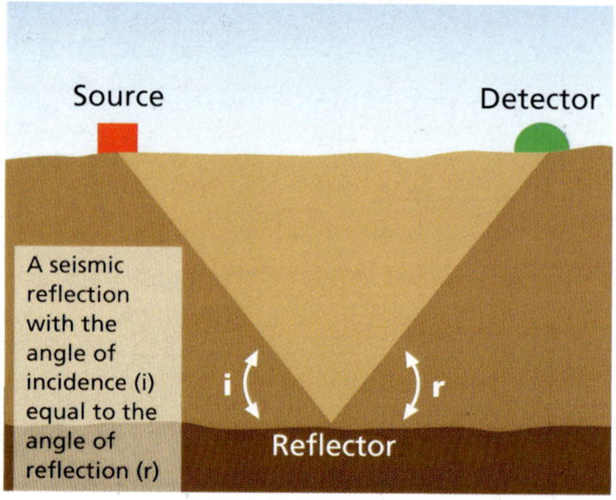

Using a vibrator truck

		S	R
1	A vibrator truck acts as the source of sound energy for a land-based seismic survey.		
2	A group of detectors called geophones is placed in a line or other pattern.		
3	Several groups of receivers are often used to form a spread.		
4	A cable is laid between the geophones and a recording truck where the data is recorded.		
5	The vibrator truck is driven to the shot point, where the sound impulses will be emitted.		
6	The correct position for the truck is where the angle of incidence equals the angle of reflection (see diagram on the right).		
7	A pad or base-plate is lowered from the truck to the surface of the ground until it carries the weight of the truck.		
8	The hydraulic motors are then activated to enable the weight of the truck to shake the ground.		
9	This continues for a period of between ten and twenty seconds, and is known as a sweep.		
10	Both unwanted noise, such as from traffic or wind, and the wanted signal, are picked up by the geophone. The two must be distinguished in the recording truck.		
11	Explosives, or a thumper truck, can also be used as a seismic source on land.		
12	At sea, obviously, a vibrator truck cannot be used. The most common alternative is an array of air guns towed behind a seismic ship.		
13	The signals are received by hydrophones connected to a cable, the streamer.		

2 Draft editing

Reread Focused reading 1 and then read the following passage.

Principles of seismology

Seismology is based on the mathematical principles of acoustics. A sound wave is emitted through a rock formation. When it reaches a particular boundary in the formation, it is reflected, and the total time it takes to reach the receiver is measured. This is called its travel time. The amplitude of the seismic echo off the top of a surface depends primarily on the contrast in acoustic impedance (sound velocity multiplied by density) between the upper and lower rock layers that form the surface. The greater the contrast, the larger the reflection. The percentage of seismic energy reflected is called the reflection coefficient. Typical sedimentary layers have a reflection coefficient of between 2 and 4%. Gas has a very slow velocity and, if the layer is overlain by a seal, the acoustic impedance contrast will produce an echo of about 16% of the seismic energy, called a bright spot. This is often an indicator of the presence of hydrocarbons. Another type of direct hydrocarbon indicator is a flat spot, which is a level, flat seismic reflector in a petroleum trap produced by rock layers that are not flat, such as an anticline.

The following sentences contain a number of factual mistakes. Amend or rewrite the sentences in the spaces provided so that they make sense.

1 The greater the impedance of the formation, the shorter the travel time of the sound wave.

2 Acoustic impedance is calculated by dividing sound velocity by density.

3 The smaller the contrast in acoustic impedance, the greater the reflection.

4 Typical sedimentary layers normally have a very high reflection coefficient.

5 Gas increases the velocity of sound passing through it.

6 It is easier to tell the difference between oil and rock than between gas and rock.

7 A bright spot is where the contrast in impedance is very low.

8 A flat spot shows a petroleum trap in a flat layer of rock.

Refresh

Refresh the vocabulary you have learnt in this unit.

Put it to work

Describe a seismic survey you have carried out. Describe the process of seismic acquisition and the results of the survey.

Section 4
Geological and Geophysical Operations

6 Mud Logging

While a well is being drilled, a production company will carry out many tests, recording their findings in well logs. One of the most common is a mud log, a chemical analysis of the mud and cuttings brought up during drilling. This unit focuses on vocabulary which will help you to describe the use and interpretation of mud log data in English.

Focused reading

1 Mud log diagram

Here is an example of a mud log diagram. Read the explanation given in the key, opposite, and label the log according to the information given, including units of measurement where appropriate.

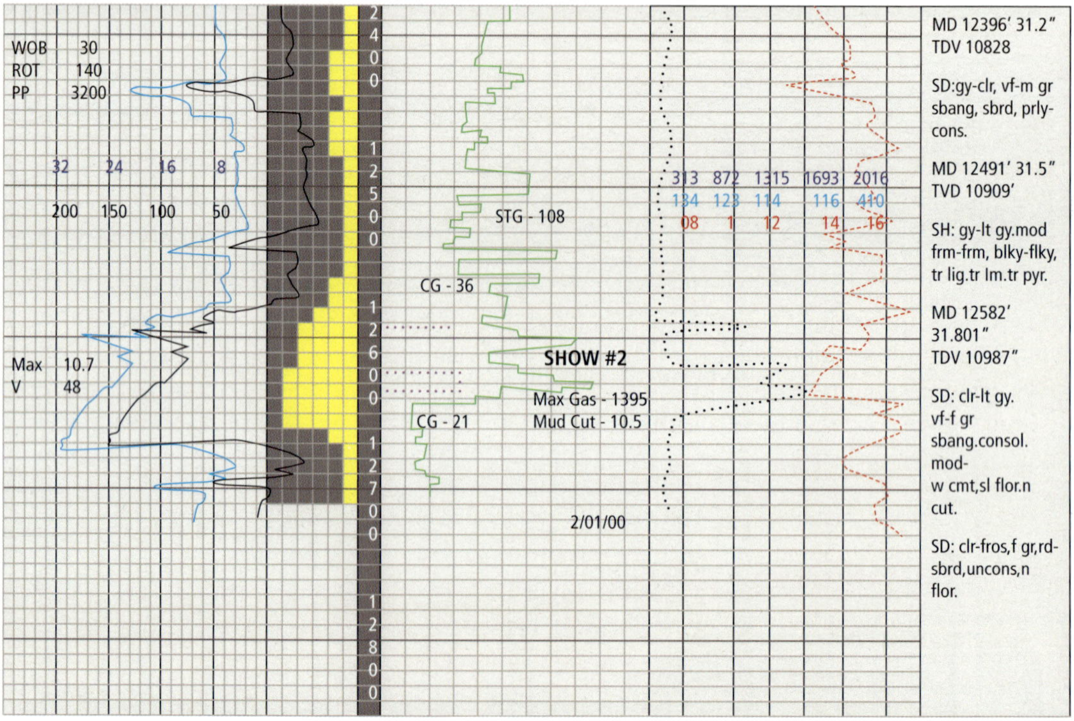

The **ROP** (rate of penetration) is represented by the black line on the left side of the log. The further to the left the line goes, the faster the rate of penetration. On this mud log, ROP is measured in feet per hour, but on some older, hand-drawn mud logs, it is measured in minutes per foot. Faster rates are an indication of softer rock formations.
The **porosity** is represented by the blue line furthest to the left of the log. It indicates the pore space within the rock structure. Notice how far to the left the porosity goes where all the sand (in yellow) is. This indicates that the sand has good porosity. Porosity is not a direct or physical measurement of the pore space, but rather an extrapolation from other drilling parameters and therefore not always reliable.
The **lithology** is represented by the cyan, grey/black and yellow blocks of color. Grey/black = shale and yellow = sand. More yellow represents more sand identified at that depth. The lithology is measured as a percentage of the total sample, as visually inspected under a microscope. These are only a fraction of the different types of formations that might be encountered.
The **gas** is represented by the green line and is measured in units of ppm (parts per million) as the quantity of total gas, but does not represent the actual quantity of oil or gas the reservoir contains. The squared-off dash-dot lines just to the right of the sand (in yellow) and left of the gas (in green) represents the heavy hydrocarbons detected.

2 Alternative systems

This is an extract from a mud logging company's website promoting their Data Acquisition System. This system presents the data in a different form to that of the mud log shown in the diagram opposite. Underline the key elements of the digital recording service described on the website.

http://www.mud-logging.com

Digital Recording

We have developed one of the most dependable and accurate Computer Data Acquisition Systems in the industry. Depth, Rate of Penetration, Total Gas and Chromatograph component gases are all read and plotted by the computer. Plotting of component gases by the computer is a revolutionary development that gives an accurate depiction of Show Zone character and provides a tremendous database to utilize Pixler Ratios to evaluate the wellbore.

For further information the DAS, visit http://www.pmles.com/live-well-data-247

Source: Top Dog Mud Logging (www.topdogmudlogging.com)

Vocabulary-building activities

1

On-site and remote mud logging

Here are two extracts from a company website advertising their mud logging services. The first describes a service that includes two technicians working on-site. The second service is managed remotely from the company's offices. Fill in the blanks using the words in the box.

compatibility   cuttings   instrumentation   interpretations   loggers   manned   remotely   reports   submitted   tours   trailer   zone of interest

http://www.mud-log.com

Two Technician Mud Logging Service

With this service, we provide the following:

- one Computerized Mud Logging **(1)** ________ equipped with Digital Data Acquisition System, Total Gas and Chromatograph;
- logging trailer **(2)** ________ 24 hours each day. Two technicians rotate **(3)** ________ and each is responsible for keeping all records and reports updated and **(4)** ________ on a timely basis;
- continuous monitoring of gases coming from the wellbore;
- lithological **(5)** ________ and fluoroscopic examination of well **(6)** ________ caught at properly lagged intervals by the loggers;
- Trip Gas Reports after each trip to help determine where gases are coming from;
- Wellsite Information Transfer Standard (WITS) Compliant;
- updated mud logs and drilling/geological **(7)** ________ are e-mailed daily.

http://www.mud-log.com

Remote Service Mud Logging

This service:

- uses our standard digital mud logging **(8)** ________ and provides: Total Gas, Chromatographic Analysis and Drilling Time (in minutes per foot);
- uses satellite technology to **(9)** ________ operate the unit;
- has WITS **(10)** ________ – i.e., gas readings can be viewed on Drilling Rig instrumentation;
- includes experienced **(11)** ________ who monitor units at least every three hours;
- includes a local technician directed by the central monitoring station;
- guarantees efficient data dissemination, whether it is across the wellsite or across the continent;
- enables clients, using Internet Explorer, to view 'zones of interest' as they are being drilled;
- includes the option, at Operator's request, for one of our experienced staff to be at the wellsite when a **(12)** ________ is being drilled (additional charge).

2 Well log terminology

Here is some other terminology associated with well logs. Use your knowledge and the information available in this unit to match the word or phrase to the correct definition.

cased-hole logging core depth strip lithologic log openhole logging rock sample sidewall coring sonde wireline well log

1 A physical description of the rocks through which the well is drilled.
2 A vertical scale of depth in the well.
3 A collection of cuttings which are examined for evidence of hydrocarbons.
4 A cylinder of rock drilled from the formation with special equipment.
5 An operation to take cuttings from a specific area of drilling, sometimes using explosive charges fired into the wall of the well.
6 Information gained by using an electric tool lowered into the well on a cable.
7 An electric tool used in wireline well logging.
8 Logging while the walls of the well are still bare.
9 Logging in wells in which casing has been cemented into the well.

3 Logging activities

Underline the verbs which describe logging activities.

assessing	attaching	calculating	capping
checking	circulating	converting	extrapolating
identifying	lending	mapping	measuring
monitoring	plotting	sampling	sealing

Refresh

Refresh the vocabulary you have learnt in this unit.

Put it to work

Choose one of the two mud logs shown in this unit. If you were a mud logger, what information would you supply to your client on the drilling rig? You can deliver your report orally or in writing.

Glossary

Unit 5

Seismic surveying	
seismic acquisition	use of acoustic techniques to acquire geological data
sound wave	the way that sound travels
sound source	a piece of equipment from which sound can be emitted
reflect the sound	throw back the sound
boundaries	borders between layers of rock
sound receiver	a piece of equipment designed to receive sound and then transmit it for recording
geophone	a sound receiver used in onshore exploration (as opposed to a hydrophone, which is used at sea)
echo	the reflection of sound waves
velocity	the speed at which sound travels
amplitude	the size of sound waves
acoustic contrast	the difference in reflection (or echo) between different layers in a formation
offset	the distance between the source and the receiver
random noise	noise which is unwanted in the survey
coherent signal	the signal which is wanted because it gives information about the formation
2D/3D/4D seismic	2D refers to two-dimensional seismic (height and length); 3D refers to three-dimensional (height, length and depth); 4D is 3D plus time
vibrator truck	a truck used for creating a sound impulse
spread	several groups of receivers arranged around the survey site
shot point	the point from which the sound is emitted
impulse	a brief emission of sound
emit	to send out (sound, light, etc.)
sweep	a continuous emission of sound over a short period
thumper truck	a truck which creates a sound impulse with a single impact on the surface of the formation
array	in marine seismic, a group of air guns towed behind a ship
tow	to pull with a vehicle, such as a car or ship
hydrophone	a sound receiver used in marine seismic
streamer	a line of air guns and hydrophones on a cable, towed behind the ship
travel time	the time taken by a sound wave to travel from the source to the receiver
impedance	resistance to the sound wave caused by the layers through which it passes
reflection coefficient	a mathematical calculation of acoustic contrast between two layers

Unit 6

Mud logging	
mud logging	acquiring data about the characteristics of a formation by examining mud from a drilling operation
well log	a report obtained by mud logging or another form of logging operation
rate of penetration (ROP)	the speed at which a drill travels through the rock formations
extrapolation of data	calculations based on a number of different sets of data, called extrapolation
lithology	the macroscopic nature of the mineral context, grain size, texture and colour of rocks
chromotograph	a piece of equipment used to show the different chemical elements in a compound
plotting	marking on a position on a graph or chart
show zone	the area of the formation where logs report evidence of hydrocarbons
pixler ratios	calculations of the ratios between the different chemical elements in a zone of interest
instrumentation	a group of measuring instruments
manned	having a crew operating the equipment (as opposed to unmanned or remote control operation)
submitted	delivered as, for example, a report or a job application
trailer	a vehicle pulled behind a truck or tractor, for example
zone of interest	the area of formation, where hydrocarbons are believed to be present
lagged	allocated a time interval at which data will be acquired (in drilling, lag is the time it takes for mud to reach the surface of the well)
cased-hole logging	logging in wells in which pipes have been cemented into the well, ready for production
openhole logging	logging while the walls of the well are still bare
core	a cylinder of rock drilled from the formation with special equipment
rock sample	a collection of cuttings which are examined for evidence of hydrocarbons
depth strip	on a log, a vertical scale of depth in the well
sidewall coring	an operation to take cuttings from a specific area of drilling, sometimes using explosive charges fired into the wall of the well
sonde	a cylinder filled with instruments attached to a wireline
wireline well log	information gained by using an electric tool lowered into the well on a cable
lithologic log	a physical description of the rocks through which the well is drilled
bare walls	walls of a well which have not been 'cased'

Section 4
Geological and Geophysical Operations

7 Wireline Well Logging

The wireline truck, which operates the wire cable used to raise and lower tools into the well, is an important feature of onshore oil and gas rigs. Well log data is used to make operational decisions about the well, to correlate formation depths with surrounding wells, and to judge the quantity and quality of hydrocarbons present. This unit focuses on vocabulary which will help you to describe the use and interpretations of wireline well log data in English.

Focused reading

1

Wireline well logging

Josef, the trainer we met in Unit 4, is continuing his training for junior field engineers, and here he is introducing the topic of wireline well logs.

1 What exactly is a sonde?
2 How does it work?

At various stages during drilling, the drillstring is raised from the hole and a sonde containing a number of different logging tools is lowered to the bottom. The sonde is attached to armoured cable and is reeled back up at a steady speed, transmitting data to the loggers on the surface. The sonde makes contact with the well sides as it ascends, by means of expandable arms or bow-springs, which also act as transmitters for the electrical, acoustic or radioactive properties being examined in the formation. The information is recorded in different tracks on the log (Tracks 1, 2 and 3, from left to right), alongside the depth strip. Zones of interest are shown by kicks when the measurement moves from one side to the other.

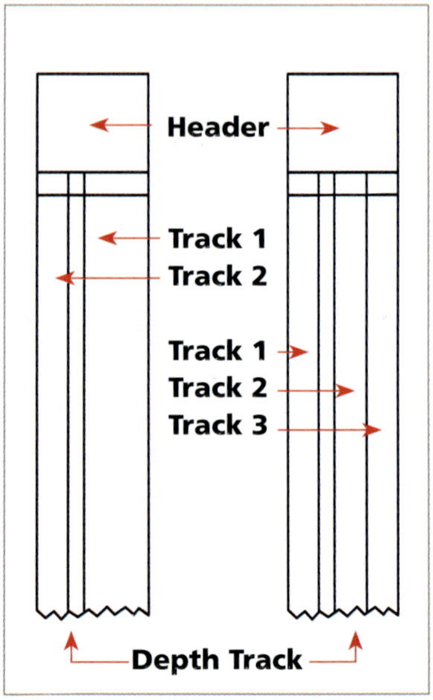

Wireline well log formats

2 Different types of wireline well log

Josef goes on to give examples of different wireline logs. Make notes as you read.

There are many different types of log – here are some of the most common examples.

The first type of log, invented by the Schlumberger brothers in the 1920s, was an Electrical Log used to measure resistivity. It passed an electrical current into the formation and the resistivity to the current indicated the presence of hydrocarbons. Nowadays, resistivity is measured in other ways, but the electrical log is still used to measure the current generated by contact between mud and natural waters in the pores of a reservoir rock. This is known as Spontaneous or Self Potential (abbreviated to SP) and is recorded on Track 1 of the log. Different types of sedimentary rocks have different SP signatures, enabling identification by the petrophysicist.

Nowadays, resistivity (or its inverse, conductivity) is measured by an Induction Log (IL), which allows the electric current to be focused and induced in a very specific zone of the formation. The tool can measure shallow, medium or deep resistivity depending on how far into the formation the current is passed. Oil and gas have very high resistivity and their presence shows a kick on the Track 2 log.

A third type of log is the Gamma Ray Log (GRL), which uses a scintillation counter to measure natural radioactivity in the rock, and is plotted on Track 1. As I'm sure you remember, there are three main sedimentary rocks – shale, limestone and sandstone. Only shales are radioactive and they kick to the right on the log. Other sedimentary rocks kick to the left. Gamma ray logs are frequently used because they can be run in both openhole and cased-hole and are relatively cheap.

Then, there are two types of log which act as radioactive sources. The first is a Neutron Porosity Log (NL), which bombards the rock formation with high speed atomic particles (neutrons). When a neutron meets a hydrogen particle, a lot of its energy is absorbed and gamma rays are produced. On the log, either the slow-moving neutrons or the gamma rays are counted, and this shows the porosity of the rock because the greater the number of neutrons and gamma rays, the more porous the rock. The log shows porosity as a percentage in Track 2.

The other is a Formation Density Log (FDL), also called a Gamma-Gamma Log. This measures the density of the subsurface rock by emitting gamma rays, which are absorbed when they encounter dense formation, but less so when they meet porous rock. So, in this case, a high measurement is an indicator of hydrocarbons, shown on Track 2 of the well log.

The last log I'm going to mention is the Caliper Log (CAL). This is used to measure the diameter of the hole. It has sensors which brush the side of the hole as it is run up and generate a record of the size, recorded on Track 1. The hole size is important for making calculations of cement quantities for casing, and also for the calibration of compensated logs.

Vocabulary-building activities

1

Sonde anatomy

Reread Focused reading 1 and label parts A, B and C on the diagram of a sonde below.

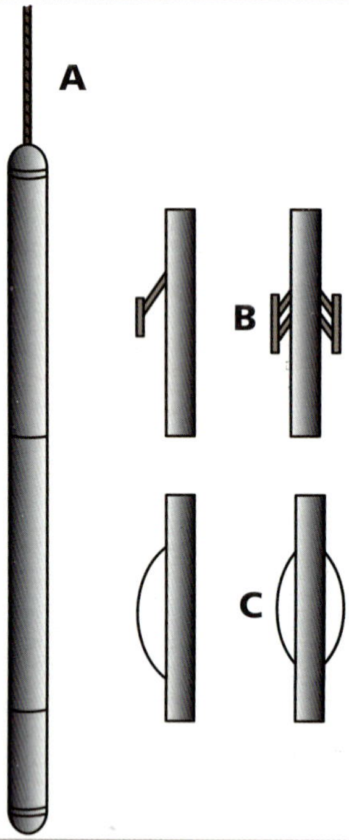

Sonde on wireline

2

Wireline logs and their uses

Fill in the table below in as much detail as you can from the information provided in Focused reading 2, on the previous page. See if you can complete the table by carrying out further research on the Internet.

Some Common Wireline Logs and their Uses

Name and abbreviation	Alternative name	Track	Uses
Spontaneous Potential (SP)			
Induction Log (IL)			
Gamma Ray Log (GRL)			
Neutron Porosity Log (NL)			
Formation Density Log (FDL)			
Caliper Log (CAL)			

3 Alternatives to wireline logging

Measurement while drilling (MWD) and logging while drilling (LWD) are alternatives to wireline drilling which allow real-time acquisition of data during a drilling operation. Here are some key facts about MWD and LWD. Match up the two parts of each sentence to form complete and factually correct sentences.

1	MWD is often used in directional drilling	a	and data is transmitted to the surface by fluid pulses in the drilling mud.
2	LWD provides an alternative to wireline drilling	b	showing the orientation of the drill bit.
3	Sensors are attached to the drillstring just above the bit	c	because it acquires information about deviation and azimuth.
4	Using a magnetometer,	d	which automatically adjusts the direction of drilling as it goes along.
5	A horizontal well can be drilled with geosteering	e	the deviation of the well in relation to the earth's magnetic field can be measured.
6	A directional log is recorded	f	as a means of getting information about rock and fluids properties, such as resistivity, porosity and radioactivity.

Refresh

Refresh the vocabulary you have learnt in this unit.

Put it to work

Make a comparison between mud logging (see Unit 6) and wireline well logging. If you were a drilling engineer, in which situations would you hire a team of mud loggers and in which situations would you hire a team of wireline well loggers? In which situations would you need both?

Section 4
Geological and Geophysical Operations

8 Anatomy of an Oilfield

Geologists and geophysicists are often required to draw on the full breadth and depth of their research and experience in order to provide a detailed description of an entire oil and gas field. This unit focuses on vocabulary which will help you to produce such a report in English.

Focused reading

1

Global view

Here is a geological description of one of the world's major oilfields.

1 Which geological period does the oilfield date from?
2 What kind of trap has created the reservoir?
3 Describe the type, source and features of the reservoir rocks.

Ghawar is an oilfield in Saudi Arabia. It is located about 100 km (62 miles) WSW from the city of Dhahran in Al-Ahsa county of the Eastern Province. Measuring 280 km × 30 km (170 miles × 19 miles), it is by far the largest conventional oilfield in the world. The field is entirely owned and operated by Saudi Aramco, the nationalized Saudi oil company. Ghawar occupies an anticline above a basement fault block dating to Carboniferous time, about 320 million years ago; Cretaceous tectonic activity, as the north-east margin of Africa began to impinge on south-west Asia, enhanced the structure. Reservoir rocks are Jurassic Arab-D limestones with exceptional porosity (as much as 35% of the rock in places), sourced from the Jurassic Hanifa formation, a marine shelf deposit of mud and lime with as much as 5% organic material (1% to 2% is considered good oil source rock). The seal is an evaporitic package of rocks including impermeable anhydrite.

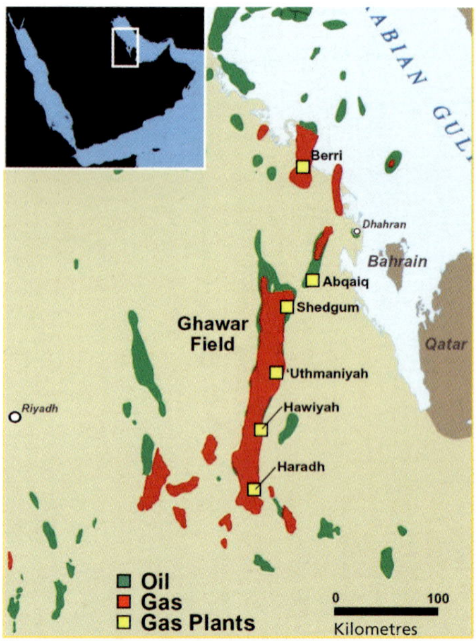

Ghawar oilfield

2 Siberian oilfield development

Here is an extract from a newspaper article about the development of an oilfield in eastern Siberia.

1 What information can you find about the geology of the oilfield?
2 Why is bringing it on-stream such a complex operation?

New Frontiers of Oil Production

TNK-BP, the Russian oil company, and the British oil company, BP, has just brought on-stream a pioneering big new oilfield in eastern Siberia. The project is part of a wave of development overcoming huge technical challenges to reach previously inaccessible crude oil reserves in Russia. The crude oil at Verkhnechonskoye is some of the oldest in the world, formed 500 million years ago from fossils of some of the first life forms on the planet. That makes it 300 million years older than the world's deepest freshwater lake, Baikal, which is the nearest well-known natural landmark.

The age of the oil deposit, which takes its name from the nearby head of the river Chona, creates special problems. For a start, there is almost no experience in the industry in dealing with such ancient pre-Cambrian rock. 'There are only one or two fields in the entire world as old as Verkhnechonskoye,' said Summers, the TNK-BP chief operating officer.

The temperature in the reservoir is about 18 degrees Celsius (64.4

Fahrenheit), much colder than in other fields, said Sergei Brezitsky, executive vice president for upstream operations at TNK-BP. 'By the time the oil reaches the surface, it is at minus 2 degrees.' Hot water must be pumped around the pipe to keep the oil flowing. The rock containing the oil is also salty, so the company is experimenting with ways to reduce the salinity to acceptable levels for pumping into the pipeline network.

Finally, the geological structure of the oil deposits is complex, meaning that with conventional drilling, a large number of expensive wells would be needed to maximize production. Soviet oil engineers drilled perhaps 100 exploratory wells at the field. They gave up because of the difficulties and lack of infrastructure.

Vocabulary-building activities

1

Fuller explanation

Reread the two Focused reading passages. The following sentences expand on or explain extracts that you will find in one or both of those passages. Identify the phrase/extract.

Example:

'A stratigraphic trap above a structural fault in the hard layer of rock below the sedimentary layer' *gives fuller explanation of the phrase* 'an anticline above a basement fault block' *in the first passage.*

1 During the period when the rock formations on the coast of north-east Africa pushed up against the south-west coast of Asia.
2 The completion of the petroleum trap consists of rocks originally created by evaporation.
3 Crude oil reserves which could not have been reached before this operation started.
4 The rock formation here is much older than nearly all other oil and gas fields in the world.
5 The rock is very salty and the company is trying to find ways to make it less so.
6 Previous efforts to exploit the area failed because of the cold temperatures, the age of the rock formation and because there were no roads.

2

Word puzzle

Find the answers to these clues and fill in the grid opposite. The key letters, which fall in the highlighted squares, form a very important measurement for oil and gas drillers.

1 A geological feature where oil and gas can be contained.
2 On a map, an indication of the distance represented by the map.
3 What hydrocarbons do when they move through a layer of sedimentary rock.
4 In seismic, an acoustic receiver used in land exploration.
5 The area of rock where petroleum originally comes from and the origin of sound in a seismic survey.
6 An example of a stratigraphic trap, a bend in a layer of sedimentary rocks.
7 A flat surface that can be shown on a map or in a mathematical figure.
8 The most common sedimentary rock, found in riverbeds and seabeds, etc.
9 A piece of electrical equipment filled with measuring devices and used in wireline logging.
10 This means that the oil and gas in a reserve is getting less and less.
11 A sedimentary rock consisting mostly of grains of sand.
12 With holes which allow hydrocarbons to migrate.
13 The dense layer of rock which prevents petroleum from escaping from a trap.
14 This describes a hill with very small gaps between the contour lines.
15 The physical characteristics of a rock formation.
16 The speed of, for example, a sound wave as it travels through rock.
17 The flow of electricity or water.

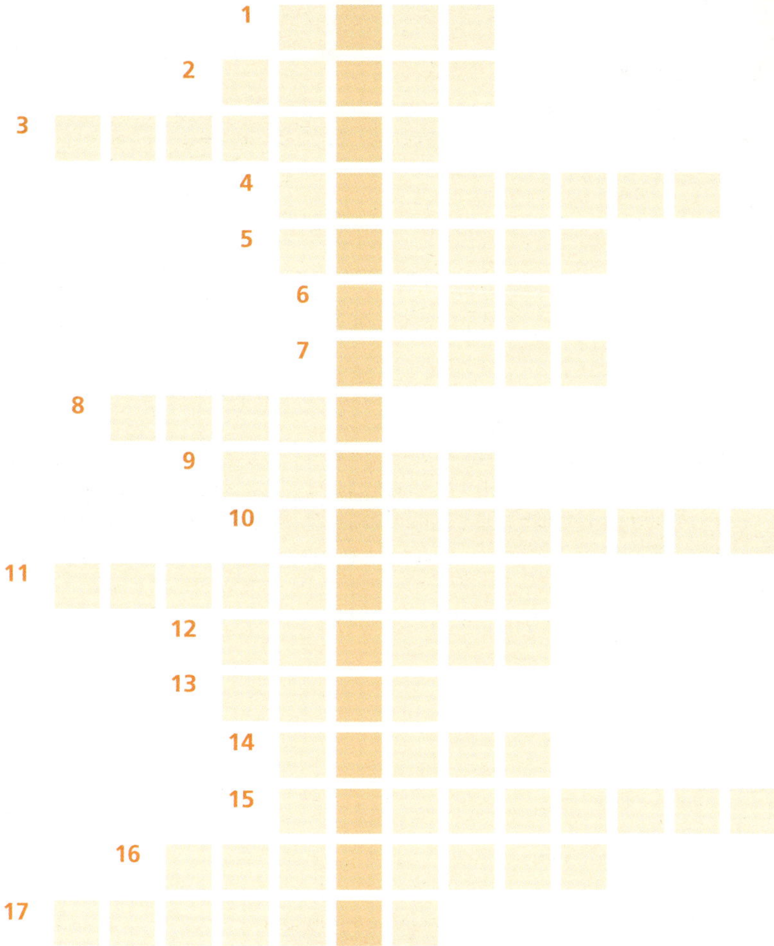

Put it to work

Prepare your own 'Anatomy of an Oil and Gas Field' on a field of your choice, researching as necessary to find as much geological and geophysical information as possible. Write out and, if possible, deliver a short presentation about the field, including information about:

- how the field was discovered and drilled;
- any important events in the development of the field;
- the lithology of the field – source, reservoirs, traps, etc.;
- any maps or cross sections available;
- any well logs available, geological or wireline;
- the future prospects for the reserve.

Refresh

Refresh the vocabulary you have learnt in this unit.

Glossary

Unit 7

Wireline well logging	
wireline	a wire rope used to raise and lower tools in a well
sonde	a cylinder filled with instruments attached to a wireline
armoured cable	a cable protected by strong outer walls
reel back up	cable or wireline is reeled (turned around a reel) into and then back up the well
expandable arms	on a sonde, the arms expand out from the sonde until they touch the well wall
bow-springs	a flexible steel bow on the side of a sonde, allows the sonde to fit the well, but still travel and transmit data
track	on a well log, data curves are represented in vertical sections called tracks
kick	when the log suddenly moves from one side to the other
resistivity	the capacity of a formation to resist electric current
spontaneous (= self) potential	a measurement of the current caused by the interaction of drilling mud and natural formation fluids
conductivity	the capacity of a formation to conduct electric current
induction log (IL)	a log that measures the resistivity of rocks by inducing an electric current
induce a current	to produce a current
gamma ray log (GRL)	a log which measures natural radioactivity in a formation
scintillation counter	a device that counts flashes of light produced by radioactive material
neutron porosity log (NL)	a radioactive log used to measure porosity
bombard	to attack continuously
neutron	a particle with no electric charge in the nucleus of an atom
formation density log (FDL)	also called a gamma-gamma log; a radioactive wireline log used to measure density of a formation
encounter dense formation	to meet a dense formation
caliper log (CAL)	a log used to measure the diameter of the well
calibration	checking a reading against a standard reading in order to check its accuracy
compensated log	a wireline log that has been adjusted to take into account irregularities of well size and roughness
measurement while drilling (MWD)	taking measurements related to the direction of drilling while drilling is in progress
logging while drilling (LWD)	taking measurements related to the characteristics of the formation while drilling is in progress
real-time acquisition	getting data at the same time as drilling
fluid pulses	movements in a rhythmic way through the drilling fluid
azimuth	the position of an object measured horizontally in degrees clockwise from magnetic north
magnetometer	a device for measuring the intensity of the earth's magnetic field
geosteering	using MWD and LWD techniques along with a steerable downhole assembly for drilling horizontally through a formation

Unit 8

This unit reviews some of the words and phrases you have seen in some of the previous units in this section. Words and phrases which are new are defined. Important items you have seen before are referenced.	
nationalized	became owned by the state – before the company was nationalized, it was a private company
anticline	see Unit 4.2
basement	unproductive rock lying below sedimentary rocks
fault	see Unit 4.2
impinge	to come up against and have an effect on
enhanced	helped to develop
porosity	see Unit 4.1
sourced	see Unit 4.1
shelf	see Unit 4.2
deposit	see Unit 4.2
lime	short for limestone (see Unit 4.1)
seal	see Unit 4.1
evaporitic	caused by evaporation
package of rocks	a collection of rocks
anhydrite	a salt mineral (CaSO4)
brought on-stream	made productive
pioneering	the first of its kind, leading other companies
a wave of development	a rapid increase in production
previously inaccessible	before this operation, it could not be accessed
fossils	the remains of a prehistoric animal or plant
landmark	something that can be seen from a big distance around
salinity	the level of salt in the formation
infrastructure	roads, railways and other means of travel and communication
stratigraphic trap	see Unit 4.2
sedimentary rocks	see Unit 4.1
acoustic receiver	see Unit 4.5
measuring devices	pieces of equipment used for measuring
grains of sand	the particles that sand is formed from
migrate	see Unit 4.1
dense	thick and difficult to penetrate
contour lines	see Unit 4.3
sound wave	see Unit 4.5

Section 5 Contents

Producing and Delivering

	Titles	Focused Reading	Vocabulary Areas	Aim(s)
1	Well Completion	• Web page selling well completion services • Online encyclopedia entry about well completion	• Completion processes and systems • Openhole and cased-hole completions • Parts of a Christmas tree	To be able to describe well completion.
2	Surface Treatment	• Extract from training material about oil and gas separators • Sales brochure and diagram of a vertical oil and gas separator	• Oil and gas separators and separation • Treatment processes • Measurement during storage and transfer	To be able to describe surface treatment.
3	Workover	• Transcripts of an interview between a journalist and an engineer	• Workover jobs and operations • Workover problems and solutions • Prepositions	To be able to describe workover problems and possible solutions.
4	Petroleum Production	• Article from a web-based magazine • Advertisement for a GasGun	• Techniques for maximizing production	To be able to describe techniques for maximizing production.
5	Improved Oil Recovery	• Extracts from a conference presentation on improved oil recovery	• Methods of enhancing oil recovery • Natural drives and processes	To be able to describe tertiary recovery methods.
6	Pipelines	• Web page selling Pipeline Security and Leak Detection System • Transcript of a meeting between a sales manager and a head of Pipeline Security	• Pipeline networks and systems • Pipeline statistics • Word partnerships with adjectives and adverbs	To be able to describe pipeline transport.
7	Tankers	• Three extracts about oil tankers from different types of source media	• Tanker transportation • Economies of scale, pricing and profits • Cost of oil spills	To be able to describe tanker transport.
8	Science in the Oil and Gas Industry	• Web page from a company offering assaying services • Web page describing different grades of crude oil	• Grades of crude oil • Scientific properties • Assaying	To be able to describe scientific work in the oil and gas industry.

Section 5
Producing and Delivering

1 Well Completion

After a well has been drilled and tested, it can either be plugged and abandoned as a dry hole, or it can be completed as a producer by setting pipe. This unit focuses on vocabulary which will help you to describe well completion in English.

Focused reading

1

Selling completion systems

BJ Services is a big oil and gas services company. Here is an extract from the company's web page about completion services (www.bjservices.com). Underline the positive phrases used to sell the service that they offer.

http://www.bjservices.com

Completion Systems

BJ Services offers a complete line of *completion systems* for all applications, including conventional completions and horizontal wells in both gravel-packed and conventional configurations. Products range from conventional Permanent Packer Systems to high-performance Retrievable Gravel Pack Systems (10,000 psi). These systems have been field tested and proven to work in the most demanding environments. BJ's tool systems are specifically designed to handle the increased demands of today's high rate, high pressure frac-pack completions.

All tool components are designed for maximum erosion resistance, allowing rates up to 35 barrels per minute at proppant volumes up to 400,000 pounds, and all are 10,000 psi differentially rated. Bonded seals are used at all critical points for positive sealing integrity.

The concept of a reliable one-trip perforating and Gravel Pack/Frac Pack system has long been the goal of operators and service companies. One of the main problems has been the extreme force generated during perforating gun detonation, which has been the cause of damaged packers, twisted screens, premature packer setting, and other types of failures. BJ Services' solution to the problem is the *ComPlete Releasable Single Trip (RST) System*.

2 Completing

Here is an extract from an online encyclopedia about well completion.

1. What is the difference between an 'openhole completion' and a 'cased-hole completion'?
2. How is the flow of hydrocarbons into the well improved?
3. What artificial ways of bringing oil and gas to the surface are used?

http://www.oilandgasdictionary.com

C Completion

After drilling and casing the well, it must be 'completed'. Completion is the process in which the well is enabled to produce oil or gas.

In a cased-hole completion, small holes called perforations are made in the portion of the casing which passed through the production zone, to provide a path for the oil to flow from the surrounding rock into the production tubing. In openhole completion, often 'sand screens' or a 'gravel pack' is installed in the last drilled, uncased reservoir section. These maintain structural integrity of the wellbore in the absence of casing, while still allowing flow from the reservoir into the wellbore. Screens also control the migration of formation sands into production tubulars and surface equipment, which can cause washouts and other problems, particularly from unconsolidated sand formations in offshore fields.

After a flow path is made, acids and fracturing fluids are pumped into the well to fracture, clean or otherwise prepare and stimulate the reservoir rock to optimally produce hydrocarbons into the wellbore. Finally, the area above the reservoir section of the well is packed off inside the casing and connected to the surface via a smaller diameter pipe called tubing. This arrangement provides a redundant barrier to leaks of hydrocarbons, as well as allowing damaged sections to be replaced. Also, the smaller diameter of the tubing produces hydrocarbons at an increased velocity in order to overcome the hydrostatic effects of heavy fluids, such as water.

In many wells, the natural pressure of the subsurface reservoir is high enough for the oil or gas to flow to the surface. However, this is not always the case, especially in depleted fields where the pressures have been lowered by other producing wells, or in low permeability oil reservoirs. Installing a smaller-diameter tubing may be enough to help the production, but artificial lift methods may also be needed. Common solutions include downhole pumps, gas lift or surface pump jacks. Many new systems in the last ten years have been introduced for well completion. Multiple packer systems with frac ports or port collars in an all-in-one system have cut completion costs and improved production, especially in the case of horizontal wells. These new systems allow casings to run into the lateral zone with proper packer/frac port placement for optimal hydrocarbon recovery.

Vocabulary-building activities

1

Cement job

There are three stages involved in a cement job:

- Preparing the well
- Running casing
- Cementing

A number of different steps are involved before each stage of the job is completed. Rearrange the steps described below into the most logical order, under each of the three stages.

1 A cementing head is attached to the wellhead to receive slurry through a line from the pumps.
2 A temperature log is run to locate the top of the setting cement from the heat it gives off.
3 A wet slurry is prepared by mixing together sacks of dry cement and water.
4 Casing string is guided down the well using a guide shoe and centralizers.
5 Cement additives, for example, to retard or accelerate setting time, are added.
6 Mud is circulated for a period of time to remove any remaining cuttings.
7 One by one, each casing joint is stabbed into the casing string already in the hole.
8 The cement is allowed to set (waiting on cement, WOC) for 8 to 12 hours.
9 The cement slurry is pumped down the hole.
10 The joints are screwed together using casing tongs.
11 The threads of the joints are sealed tight with a thread compound.
12 The well is conditioned by running a drillstring with a used bit into the well.
13 The wiper plugs, guide shoe and any cement at the bottom are drilled out.
14 Two wiper plugs are run into the hole to force the slurry through the guide shoe and up into the annulus of the well.
15 Wall scratchers are run up and down or rotated in the well to scrape filter cake off the well sides.

2

Openhole or cased-hole completions?

Reread Focused reading 2. Do you think these sentences refer to openhole or cased-hole completions?

1 There is no casing in the producing formation.
2 It is more expensive because more casing is used.
3 It can only be used where the formation is clearly defined.
4 It cannot be used in soft formations which might cave into the well.
5 The casing is set before the pay is drilled.
6 In sand formations, an underreamer is run to make a cavity in the pay zone.
7 A gravel pack can be used to consolidate the soft formation.
8 Fluids can flow through a screen or slotted liner in the well.
9 Perforations are shot through the casing and cement and into the formation.

3 Christmas tree

Use the information given in the paragraph below to label the diagram of a Christmas tree. Only use the words which are highlighted for your labels.

Oil wells where oil can flow without artificial help (and gas wells which always flow naturally) are fitted with a *Christmas tree* – a series of pipes, fittings, valves and gauges. The Christmas tree is attached to the *wellhead*, the large steel fitting on top of the well, consisting of a *tubing head* on top of the *casinghead*. Sticking out of the lower section of the Christmas tree is the *master valve*, to turn off the well in an emergency. Above that, there is one, or sometimes two, wings, depending on the number of producing zones being controlled. Each *wing* has a *valve* to regulate flow. At the top of the Christmas tree is a *pressure gauge* to measure tubing pressure.

A Christmas Tree

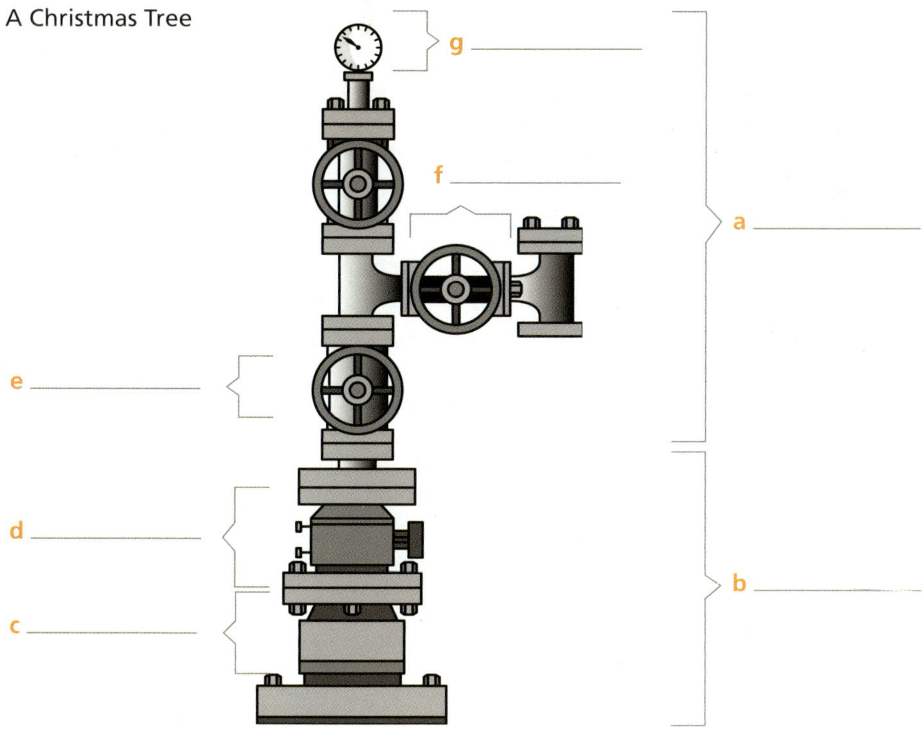

Refresh

Using one of the strategies for consolidating and storing new vocabulary recommended in the Introduction, refresh what you have learnt in this unit.

Put it to work

You have been asked to assist a trainer working on a course for trainees in oil and gas production. He has asked you to design a PowerPoint slide entitled 'Key Elements of a Completion System', comprising a list of the key elements and a short explanation of each, listed with bullet points.

2 Surface Treatment

Once hydrocarbons have been brought to the surface, the fluid needs to be separated into oil, gas and water, and then stored. Regular measurements need to be taken during this phase. This unit focuses on vocabulary which will help you to describe surface treatment in English.

Focused reading

1

Separators and separation

Read this extract describing onshore separators from education and training material written by the training department of an oil and gas company. What are the main ways of categorizing oil and gas separators? What are the key factors involved in separation?

Oil and Gas Separators

An oil and gas separator is used to separate oil, gas and water from the stream of fluids produced by a petroleum well. It can be either vertical or horizontal and can be either spherical or cylindrical in shape.

Separators can be divided into two types – two-phase and three-phase. The two-phase type handles only oil and gas, whereas the three-phase type deals with oil, gas and water. This type is commonly called 'free-water knockout'.

In addition, a separator can be categorized by operating pressure. Low-pressure separators operate from 10 to 180 psi (69 to 1241 kPa). Medium-pressure units handle pressures of 230 to 700 psi (1586 to 4826 kPa). High-pressure equipment can cope with pressures of 975 to 1500 psi (6722 to 10,342 kPa).

The separation is mainly achieved by the force of gravity, which means that the heaviest fluid settles to the bottom of the vessel and the lightest rises to the top. The degree of separation achieved will depend on the operating pressure of the separator, on the amount of time the fluid mixture spends in the vessel and on the flow rate of the fluids.

2 Primary and secondary separation

This is an example of a vertical oil and gas separator sold by the oil and gas engineering company NATCO. Read the extract from their sales brochure, including the diagram, and explain the difference between primary and secondary separation in the Whirlyscrub V™.

Compact Porta-Test Whirlyscrub V™ for Efficient Gas/Liquid Separation

This compact, lightweight, vertical separator is ideally suited for separation processes containing moderate amounts of liquid (less than 500 barrels per MMscf). NATCO's Porta-Test Whirlyscrub V™ Separator captures over 99.9% of all liquid droplets larger than 10 microns, exceeding the performance of most vane-type separators. Over 35 years of experience with this product allows us to optimize the separator design for your application.

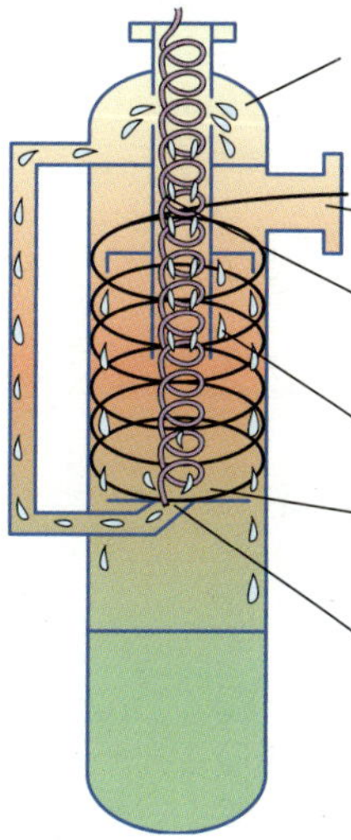

How It Works

Primary separation takes place as gas enters through a tangential nozzle, creating centrifugal force and forcing the heavier liquid particles to the vessel wall. From there, the liquids drain to the stilled chamber in the bottom of the vessel. Secondary separation occurs as the spinning gas converges at the centre of the separator and enters the vortex finder tube. Inside the vortex finder tube, the gas spins at a higher velocity and forces any remaining liquid to the tube wall. This liquid is swept upward toward the gas outlet. Prior to exiting the vessel, the liquid and a 10% side stream of gas are drawn through a small gap in the vortex finder tube and returned to the primary separation section. A low pressure area in the primary separation section created by the spinning gas provides the necessary differential pressure driving force.

Consider the Advantages

- Minimize footprint area and weight in new separator installations.
- Improve separation efficiency as gas flow rates increase.
- Operate foam-free and eliminate or reduce expensive de-foaming chemicals.

Source: Cameron System Limited

Vocabulary-building activities

1

Alternative word choices

Choose the best word to replace the underlined word or words in each sentence.

1 Every separator has a diffusion section that makes an initial separation of gas and liquid from the <u>entrance to the vessel.</u>
a) income b) inlet c) incline d) intend

2 Liquid <u>drops</u> to the bottom of the separator where the remaining gas is removed.
a) lowers b) reduces c) falls d) declines

3 Emulsion has droplets of liquid that are <u>held</u> in oil.
a) suspended b) depended c) upended d) pretended

4 In the diffuser section, fluids are <u>circulated rapidly</u> around the shell of the vessel.
a) circled b) rounded c) spun d) rotated

5 Mist extractors are used to coalesce and <u>take out</u> liquid droplets before the gas flows out.
a) replace b) replenish c) restore d) remove

6 A back-pressure valve <u>keeps at the correct level</u> the gas pressure in the separator.
a) retains b) sustains c) detains d) maintains

7 In a free-water knockout separator, water is <u>taken</u> from the bottom, oil from the middle and gas from the top.
a) drawn b) thrown c) blown d) grown

8 During the process of separating gas from liquid, liquid can be <u>soaked up</u> by silica or other suitable chemicals.
a) drained b) absorbed c) condensed d) evaporated

2

Process definition

Here are seven processes which are carried out during the treatment of gas. Which definition corresponds to each process?

absorption compression conditioning cooling dehydration stripping sweetening

1 The removal of impurities (in general) from natural gas.

2 The removal of liquids (in general) from natural gas.

3 The removal of water from natural gas.

4 The removal of corrosive gases, such as CO_2 or H_2S, from natural gas.

5 Increasing pressure on natural gas, thereby reducing its volume.

6 Lowering the temperature of natural gas to remove hydrocarbon liquids.

7 Using special chemical products to soak up hydrocarbon liquids.

3

Measurement during storage and transfer

Oil and gas are measured using a variety of different methods during storage and transfer on the surface at the wellhead. Read the passage below and make notes on the use of each of the following methods:

- stock tanks
- a gauge tape
- a shake-out test
- turbine meter
- charcoal tests
- fractional analysis
- an orifice meter

Once the oil has been separated, it goes through a flowline to the stock tanks for storage, ready for transportation to the refinery by truck, tanker or pipeline. A battery of stock tanks holds a minimum of four days' production. A gauger measures the amount of oil in the stock tanks by lowering a gauge tape with a brass weight on the end into the tank. Samples of the oil are taken and the temperature of the oil is measured. It is also important to know the amount of impurities and this is done with a shake-out test. The sample is centrifuged in a glass container and, because the water and impurities are heavier than the oil, they fall to the bottom. The volume of oil being transferred can be measured using a turbine meter. The flow of oil turns the turbine at a steady rate and the number of turns is counted. For gas, compression or charcoal tests are used to measure the condensate content of the gas. Fractional analysis produces various measurements, including relative percentage of hydrocarbon and water. Gas volume is measured by passing the gas through an orifice – a small hole – and measuring the difference between the pressures on the two sides of the orifice.

Refresh

Refresh the vocabulary you have learnt in this unit.

Put it to work

Imagine you are visiting a producing well. You are at the surface of the well. Draw a diagram of what you see and explain what you have included. Explain the purpose and functioning of any equipment and personnel you have included.

Glossary

Unit 1

Well completion	
plugged	when a well is abandoned as a dry hole and closed off with cement in the hole, it is plugged
set pipe	to place pipe in the hole
gravel-packed	when a well is completed using gravel instead of casing in the formation zone, it is gravel-packed
frac pack	a system used for fracturing wells during completion
proppant	spherical object used for keeping open fractures in the formation
bonded seal	the seal is attached tightly between the casing and the well wall
perforating	making holes (or 'perforations') in the casing
detonation	making an explosion with a spark or some other means
sand screen	a screen for preventing sand from falling into the hole
washout	a flood of water in the hole
unconsolidated	when a hole is not treated in any way, it is unconsolidated
stimulate	to assist the oil to flow more easily from the formation to the producing well
hydrostatic effects	for example, the pressure created by water in the formation
cement	a substance made of lime and clay, mixed with water
slurry	wet cement
guide shoe	a metal cylinder run on the end of a casing string to guide the string in the hole
centralizer	an attachment to the outside of the casing string to keep the string central in the well
retard	slow down
accelerate	speed up
stabbed	when one joint of a string is placed inside another, it is stabbed
set	to become hard; cement 'sets' after an hour or two
casing	steel pipe run into the well and cemented to the wall to form a producing well
tongs	a device suspended above the drill floor by cable used to hold the drillpipe or casing as it is being screwed together
wiper plugs	two cylinders made of aluminium or rubber pumped down the casing during a cement job
wall scratcher	a device attached to the outside of a casing string used to clean mud cake off the well walls
filter cake	a solid mass of clay particles on the well wall built up during the circulation of drilling mud
cave in	when the well wall collapses, it caves into the hole
underreamer	a tool run on a drillstring to make a cavity at the bottom of the well
slotted liner	a section of pipe set at the bottom of a well, which allows hydrocarbons, but prevents sediments, from flowing into the well
Christmas tree	a set of fittings, valves and gauges attached to the wellhead to control flow from a producing well
wellhead	see diagram on page 179
casinghead	
tubing head	

Unit 2

Separating	
separator	a machine used to separate hydrocarbons
free-water knockout	a process used to separate water from the hydrocarbons
settles to the bottom	falls to the bottom and stays there
compact	small and simple
capture	to take and separate from the fluids
droplets	small drops
vane	a rotating blade, part of a turbine for turning liquid, etc.
tangential nozzle	a nozzle coming in from the side of the separator
centrifugal force	a force caused by rapid turning
drain	when liquid is allowed to leave a vessel, it drains from the vessel
stilled chamber	a space at the bottom of the separator for storing the separated water
spin	to turn very quickly
converges	to come together and mix together
vortex	a phenomenon caused by rapid turning
swept upwards	carried very fast upwards
foam-free	without foam, the chemical mass caused by agitation of liquid
throughput	the amount that the separator is used
mist	a very fine cloud of vapour from liquid, with very small droplets
diffusion	spreading more widely
coalesce	come together to form a single mass
absorbed	or 'soaked up'; taken in by a chemical process
stripping	removing liquid from a gas
sweetening	removing acid gases (for example, carbon dioxide and hydrogen sulphide) from natural gas
impurities	unwanted matter
corrosive gases	gases which cause damage or destruction to metal
Measurement, testing, storage and transfer	
stock tanks	tanks where oil is stored on the surface before transportation
gauger	a person who measures the quantity of oil in a stock tank
gauge tape	a tape on a reel used to measure the quantity of oil in a stock tank
samples of the oil	small quantities used to show what all the oil is like
shake-out test	a method used to determine the amount of sediment and water in oil by applying centrifugal force
charcoal test	a test used to measure the amount of condensate in a gas (charcoal is a porous block solid produced by heating wood in the absence of air)
condensate	liquid collected by condensation, when humid air comes into contact with a cold surface
fractional analysis	a chromatographic method of testing gas for various measurements
orifice	a small opening in a skin or membrane

Section 5
Producing and Delivering

3 Workover

For optimal efficiency during production, the operator needs to monitor and maintain the flow of oil or gas at the right level. There are many different problems which may occur and which require repair or maintenance of the rig. This is known as well workover. This unit focuses on vocabulary which will help you to describe well workover in English.

Focused reading

1 Investigating workover operations

Alan Johnson, the journalist we met previously in Section 2, Unit 7 and Section 3, Unit 6, when he was researching North Sea oil and gas platforms, is now in California, visiting an oil and gas field there. He is talking to an engineer from ExxonMobil, Orville Svaranski, who is showing him a producing well.

Read the following transcript of their discussions and study the photographs, opposite.

OS: As you can see, this well is shut in at the moment.
AJ: What does that mean?
OS: Well, from time to time, a producing well needs to be maintained – we call it workover. The well is killed and we bring in a special service company.
AJ: That's their truck, is it?
OS: That's right. And the rig is theirs, too. It's a special workover rig with a mast that can be hoisted during the workover. This one is using regular tubing, but sometimes they use a coiled tube unit, which is quicker, but a bit more expensive for us!
AJ: But you're losing production while it's going on, I guess. It can't be too slow.
OS: You're right, we have to get the right balance.
AJ: It doesn't look very stable.
OS: Oh, it's certainly stable enough. The truck has outriggers to support it and those guy lines keep the mast upright. We've been using this company for twenty years now and I haven't seen one fall over yet!

1 Which of the pictures shows a drilling rig and which shows a workover rig?
2 What is the difference between the two rigs?

a

b

Workover Rig on the Acock/Anaqua Operating Co., LP #7 Scanio Shelton well in Refugio County, Texas

2 Workover jobs

Alan and Orville's conversation moves on to the subject of the different jobs that need to be done in the course of a workover.

1 Can you identify four different workover jobs that are mentioned?
2 Why is each job necessary?

AJ: So, what kind of work do they need to do during a workover?
OS: Often, you get a build-up of sand in the wellbore and that has to be removed. You can do that by pumping salt water down the annulus or by running a bailer on a wireline. That's called a sand-control job. Sometimes, the casing gets damaged by abrasion or corrosion and that can make holes or splits in the lining.
AJ: Do you have to replace all the casing then?
OS: No, fortunately there are various solutions to avoid all that hassle and expense. You can do a cement squeeze job ...
AJ: What's that then?
OS: That's when you fill gaps in the cement behind the casing. The gaps are called holidays and you have to isolate the zone using packers and then perforate the casing in that zone. Then you can pump the cement under pressure through the perforations and into the holiday.
AJ: Sounds good. Does any of the pumping equipment ever break?
OS: Yes, now and again, the pumping rod can wear out and the lower part can fall into the well.
AJ: So I guess you then have to use a fishing tool of some kind.
OS: That's right. One's called a mousetrap! And then there's sidetracking.
AJ: Sidetracking?
OS: Yeah, that's when there's something blocking the well that can't be removed or repaired. It has to be bypassed with a directional hole. A window is made in the casing just above the blockage and the well is plugged just below the window. Then, a special directional tool is used to drill off at an angle from the main well. Then you can complete the hole in the normal way.
AJ: I think I've got that. Anything else?
OS: Well, there are other workover jobs, but I reckon those are the most common.
AJ: Thanks a lot, Orville.

Vocabulary-building activities

1 Problem-solving

Here is a list of workover problems that can occur and a number of suggestions for how they might be solved. Match the problems with the solutions.

1 A lot of sand is clogging the bottom of the well.
2 The casing has collapsed.
3 The pumping rod has broken in half and a piece has fallen downhole.
4 The well has fallen in and the tube is blocked. We can't get the tube out.
5 The tubing is clogged with wax.
6 The seals on the downhole pump are worn.
7 There are gaps in the cement behind the casing.
8 There's a lot of water and mud in the well.

a Can we get it off with a paraffin knife?
b Do you think we should do a swab job?
c That means sidetracking, doesn't it?
d They'll need replacing as soon as possible then.
e We could either pump salt water down the annulus or we could stabilize it with an epoxy resin.
f We'll have to call in the fishing crew.
g We'll need to open it with a casing roller.
h I reckon squeeze cementing is the best solution for that, don't you?

2 Swabbing

In the discussion in Focused reading 2, on the previous page, Orville Svaranski does not mention swabbing – another important workover job involving the removal of mud from the well. Read this paragraph about swabbing and fill in the blanks with words from the box below.

attached	completed	hollow	killed	lower	piston	pressure	raised
removal	restore	retain	wireline				

Swabbing is the **(1)** ________ of water or drilling mud from a well, so that the oil and gas can flow into the well. A swab job can be done both after the well is **(2)** ________ to remove the last of the drilling mud, or to **(3)** ________ production in a producing well. A truck-mounted swabbing unit with a short mast is used to **(4)** ________ a swab tool down the tubing string on a **(5)** ________. A swab tool is a **(6)** ________ steel rod with rubber swab caps. When the swab tool is **(7)** ________, the swab caps seal against the tubing to act as a **(8)** ________ and lift (swab out) the liquid from the well. A lubricator (a length of casing or tubing) is temporarily **(9)** ________ above the valve on the tubing head or casinghead to provide a pressure seal. The swab tool can be run into the well under **(10)** ________ through the lubricator, so the well doesn't have to be **(11)** ________ during swabbing. An oil saver is used on top of the lubricator to **(12)** ________ any oil coming up on the wireline.

3 Workover tools

These pictures show three different workover tools.

 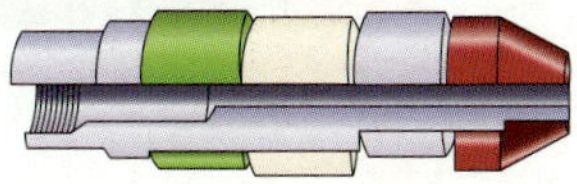 

Complete the sentences using these phrases:

scale and debris | obstacles | excess material | collapsed casing

1. The first tool is an external milling shoe – for cutting and removing ________ from equipment or tools downhole.
2. The second is a roller swage – for opening up ________ and removing ________ from the well.
3. The third is a casing scraper – for removing ________ from the internal surface of a casing string.

4 Joined-up thinking

Complete these sentences by adding suitable prepositions.

1. ________ workover operations, a rig is mounted ________ a truck or trailer ________ the well and uses a winch driven ________ the truck engine.
2. ________ stability, guy wires are attached ________ the mast and ________ anchors ________ the ground.
3. ________ a coiled tubing unit, the tubing is wound ________ a reel. It is unwound ________ the reel and lowered ________ the well.
4. The well is killed ________ filling it ________ kill fluid, pumped ________ the casing tube annulus ________ pressure ________ the tubing string is full.

Refresh

Refresh the vocabulary you have learnt in this unit.

Put it to work

Imagine that you are an engineer on a production site. One of your wells is not producing as efficiently as you would expect. List the possible reasons why this might be the case. Explain each potential problem and the workover solution you would use, with appropriate examples from your experience.

Section 5
Producing and Delivering

4 Petroleum Production

While natural pressures force oil and gas into a well and up to the surface once the well is tapped (if not before), production is often assisted by a variety of artificial means, including well stimulation, sidetracking, fracturing and acidizing. This unit focuses on vocabulary which will help you to describe these technical processes in English.

Focused reading

1

Enabling production

Here is an article from *OilVoice*, a web-based magazine for the oil and gas industry. Can you identify three techniques that were used in Orenburg to produce oil?

OilVoice

ALADDIN OIL & GAS ANNOUNCES
Successful Well Stimulation in Orenburg, Russia

Aladdin Oil & Gas Company ASA has completed the well stimulation on Well #51 at the Bogdanovskaya license in Orenburg. The well is now in production, producing 100 barrels of oil per day. Through the use of pumps, the production is expected to increase further.

The work is conducted by Schlumberger's Russian subsidiary, Petroalliance, and is completed according to the stimulation programme developed by the consultancy TRACS. The TRACS programme is providing the expected results, about 100–150 barrels per stimulated well.

Petroalliance will immediately initiate the stimulation of Well #39, and thereafter go to Well #41 and #33, as the equipment is within reach in the Orenburg region throughout the spring. Simultaneously, Well #65 will be reactivated. Subsequently, sidetrack drilling of Well #34 and #47 will be commenced. The sidetrack drilling is expected to give somewhat higher production rates, compared to ordinary well stimulation.

Aladdin Oil & Gas Company reiterates the production target of 700–1,000 barrels per day for the existing wells.

2

The GasGun

Here is an advertisement for a GasGun. Which of the following functions does the GasGun fulfil? (Three of the operations are not actually fulfilled by the GasGun, though they do have a role in preparing for them.)

well stimulation   multiple fracturing   hydraulic fracturing   well clean-up acidizing   enhanced production   waterflood

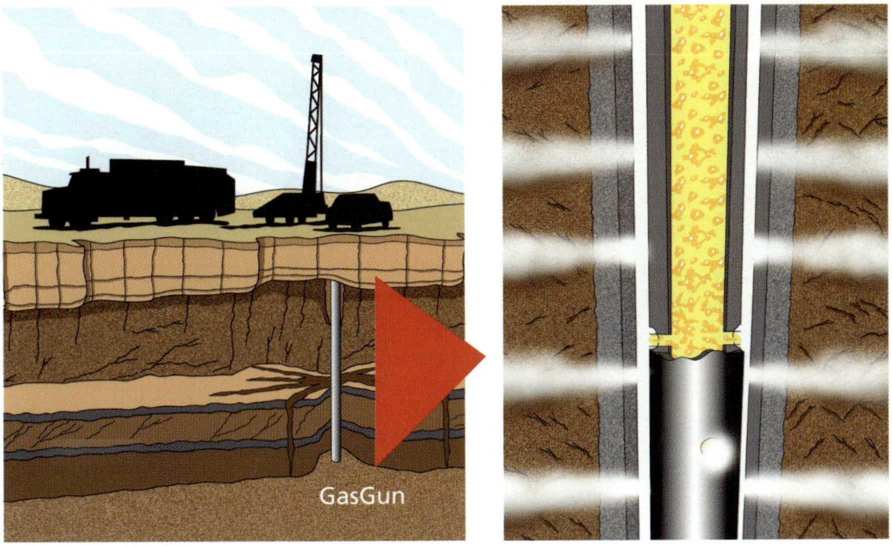

Source: The GasGun, Inc.

How it works

- Creates multiple radial fractures extending 10 to 50 feet from the wellbore. Minimal vertical growth avoids problems often associated with hydraulic fracturing.
- Removes skin and cleans up the wellbore damaged by perforators, drilling fines, cement, paraffin, mud cake, etc.
- Stimulates heterogeneous production zones, such as lenticular sands, with a higher probability of success. Fractures reach out in several directions, thus increasing your chance of intersecting the producing formations.
- Improves effectiveness of acidizing by fracturing first with the GasGun, allowing acid to etch new channels into formation.
- Enhances production in naturally fractured reservoirs by intersecting more fractures.
- Prepares well for hydraulic fracturing by breaking down formation first with the GasGun. Treating pressures are often dramatically reduced.
- Increases injection and withdrawal rates in gas storage wells damaged by compressor oils.
- Improves waterflood efficiency by providing increased and more uniform injection rates.

Vocabulary-building activities

1

Step by step

The following step-by-step description of a 'frac job', or fracture stimulation of a well, has been rearranged so that it is out of sequence. Read each step carefully and see if you can put them into the correct order.

Frac Job or Fracture Stimulation

1 The well is then reopened and is ready to test.

2 There is a substantial amount of effort spent designing the correct type of frac job for a well. The cost to perform a frac job ranges from $50,000 to more than $1 million, depending on the size and complexity of the job.

3 These fractures are extended a long distance from the well by pumping this proppant-loaded frac fluid at a high enough pressure to continue cracking the rock.

4 This crack is then filled with sand or an artificial proppant (ceramic beads, sintered bauxite or glass beads) to allow the freer flow of oil or gas from the rock to the wellbore.

5 This is a technique used by engineers to create a crack or fracture in the rock that contains oil or gas.

6 This process is carried out using trucks equipped with high-pressure pumps and sand/proppant mixers (called 'blenders') mixing the proppant with an engineered 'frac fluid' (usually a slimy looking watery substance) that is pumped down the well until the rock actually cracks or fractures.

7 When the process is complete, the well is closed and the fluid and proppant allowed to settle.

2

Acid simulation

Acidizing is a process which is sometimes used instead of fracturing as a means of well stimulation. Complete the sentences opposite by adding the clauses given in the box.

that have high temperatures –250 degrees Fahrenheit or more
a strong acid that works best on carbonate rocks
high enough to cause it to fracture
which is used mainly in limestone formations
Regardless of the acid used
known as acidizing
in which hydrocarbons and other fluids occur
that block reservoir permeability
which reacts with quartz, sand and clay
when put into contact with suitable reservoir rocks,
keeping the injection pressure low enough to avoid fracturing the formation

1 Certain acids, ________, etch or dissolve them.

2 The fact that acid reacts with rock forms the basis of a stimulation method ________.

3 Two types of oil well acidizing are fracture acidizing and matrix acidizing. Fracture acidizing involves pumping acid into the reservoir at a pressure ________.

4 In fracture acidizing, ________, the acid etches the face (each side) of the fracture to create flow channels for reservoir fluids.

5 Acid is injected down the tubing and into the permeable channels of the reservoir, ________.

6 Certain acids can remove substances ________, especially mud solids.

7 Many different kinds of acids are used in acidizing oil wells. One of the more common is hydrochloric acid, ________.

8 Acetic acid and formic acid are sometimes used to acidize carbonate reservoirs ________.

9 Another acid used is hydrofluoric or mud acid, ________.

10 ________, the idea behind acidizing is to enable reservoir fluids to flow out of the reservoir and into the wellbore

3

More or less?

Which of these verbs can be used to complete the following sentence?

allowed avoided cut damaged enhanced improved increased inhibited performed provided reached reduced removed stimulated

We have ________ production by around 150 barrels per day.

Put it to work

You have been asked to explain to a group of oil and gas trainees how oil is produced from a well. Choose three of the techniques mentioned in this unit and prepare a short presentation slide about each one. Include examples of producing wells from your own experience where possible.

Refresh

Refresh the vocabulary you have learnt in this unit.

Glossary

Unit 3

Workover	
workover	work done on a completed or already producing well to maintain, restore or improve production
producing well	a well from which oil or gas is being taken
shut in	when production on a well has stopped for maintenance
killed	when the flow of oil in a producing well is stopped; achieved by the use of kill fluid
outriggers	structures on the side of the truck to improve stability when using a workover rig
guy lines	ropes or cables used to support the workover rig
build-up of sand	a gradual accumulation of sand
bailer	a device used for removing built-up sand from the well
sand-control job	a job where built-up sand is removed from the well, or stabilized
abrasion	damage caused by friction
corrosion	damage caused by chemical reaction
splits	breaks in the lining
lining	another word for casing, the metal pipe cemented to the walls of a producing well
cement squeeze job	a job where cement is forced under pressure through holes in the casing created by perforation
holiday	a gap in the cement between the casing and the cell wall
packers	a cylinder made of rubber used to seal the casing to the well wall
sidetracking	a directional drilling technique used to avoid a blockage in the well
bypassed	avoided by drilling around the damaged section
clogging	to cause blockage by accumulating
paraffin knife	a tool used to remove build-ups of wax from a tubing string
swab job	a job in which liquids are removed from a producing well
stabilize	to make stable; to fix firmly
epoxy resin	a type of glue used to stabilize formation
casing roller	a tool used to reshape damaged casing
piston	a piece of equipment where a cylinder is moved up and down inside a tube to create a pumping movement
milling	the action of cutting or shaping metal using a rotating device
obstacles	things which block the hole and prevent free movement down and up
scraper	a tool which removes unwanted material from the surface of the casing
scale	a film of material over a flat surface
debris	another word for junk – unwanted objects in the hole
winch	a mechanical device used for lifting

Unit 4

Well stimulation	
stimulation	a process during which the flow of oil in the reservoir is activated
reactivated	when a well is opened and production restarted
fracturing	a stimulation process involving the creation of fractures in the formation along which oil can flow
acidizing	a stimulation process using acids to dissolve limestone in the formation and thus stimulate flow of oil and gas
enhanced production	extending the production life of a well using a variety of techniques
waterflood	using water in large quantities to pressurize oil towards the producing wells
radial fractures	fractures coming out into the formation in all directions from the well
paraffin	a member of the hydrocarbon series of molecules (becomes wax at low temperatures)
mud cake	a build-up of mud on the well wall
heterogenous	of the same type
intersecting	crossing over one another
etch	to cut (channels into the formation)
compressor	a machine used for reducing the volume of a gas
uniform	standardized (injection rates)
ceramic beads	small spherical objects made of ceramic
sintered bauxite	bauxite (aluminium ore) which has become a solid mass because of heating
slimy	feeling moist, soft and slippery

Section 5
Producing and Delivering

5 Improved Oil Recovery

The employment of techniques such as well stimulation and fracturization is considered to be a secondary stage of oil recovery. However, oil companies can go even further than this and use enhanced or tertiary recovery methods to draw even more of the reservoir's oil to the surface. This unit focuses on vocabulary which will help you to describe tertiary recovery methods in English.

Focused reading

Carter Maythorn is director of the Office of Fossil Energy at the Department of Energy in Washington, DC. Here, he is speaking at the 'Oil and Gas Futures' conference in Houston, Texas, about a DOE programme that is underway to enable enhanced recovery of the country's 'stranded' oil resources.

1 Primary, secondary and tertiary recovery

Read the following extract from Carter's presentation.

1 What is the difference between primary, secondary and tertiary recovery?
2 Why are US producers using tertiary recovery now?

So, what exactly is the purpose of enhanced recovery? Well, as I'm sure you know, crude oil development and production in US oil reservoirs can include up to three distinct phases: primary, secondary and tertiary (or enhanced) recovery. During primary recovery, the natural pressure of the reservoir or gravity drives oil into the wellbore, combining with artificial lift techniques (such as pumps) to bring the oil to the surface. But only about 10% of a reservoir's original oil in place is typically produced during primary recovery. Secondary recovery techniques add to the field's productive life by injecting water to displace the oil and drive it to a production wellbore, resulting in the recovery of another 20 to 40% of the original oil in place. However, a lot of the easy-to-produce oil has already been recovered from US oilfields, and so producers are increasingly using tertiary or enhanced oil recovery (also called EOR) techniques that offer us prospects for ultimately producing 30 to 60% or more of a reservoir's original oil in place.

2 Enhanced recovery methods

In this extract, Carter describes three methods of enhanced recovery. Read the extract and make notes about how each method works.

Three major categories of EOR have been found to be commercially successful in varying degrees. First of all, there's thermal recovery, which involves the introduction of heat, such as the injection of steam to lower the viscosity or thin the heavy viscous oil, and improve its ability to flow through the reservoir. Thermal techniques account for over 50% of US EOR production, primarily in California.

Then there's gas injection, which uses gases such as natural gas, nitrogen or carbon dioxide that expand in a reservoir to push additional oil to a production wellbore, or other gases that dissolve in the oil to lower its viscosity and improve its flow rate. Gas injection accounts for nearly 50% of EOR production in the United States.

And thirdly, there's chemical injection, which can involve the use of long-chained molecules called polymers. These increase the effectiveness of waterfloods, or the use of detergent-like surfactants, to help lower the surface tension that often prevents oil droplets from moving through a reservoir. Chemical techniques account for less than 1% of US EOR production. However, these techniques have been hampered by their relatively high cost and, in some cases, by the unpredictability of their effectiveness.

3 CO_2 gas injection

In this extract, Carter describes three projects in which CO_2 injection is being used to enhance oil recovery. Read the extract and make notes on each of the projects.

The EOR technique that is attracting the most new market interest is carbon dioxide (CO_2) EOR. New technologies are being developed to produce CO_2 from industrial applications, such as natural gas processing, fertilizer, ethanol and hydrogen plants in locations where naturally occurring reservoirs are not available. One demonstration at the Dakota Gasification Company's plant in Beulah, North Dakota, is producing CO_2 and delivering it by a new 204-mile pipeline to the Weyburn oilfield in Saskatchewan, Canada. Encana, the field's operator, is injecting the CO_2 to extend the field's productive life, hoping to add another 25 years and as much as 130 million barrels of oil that might otherwise have been abandoned.

We are funding another project in the Hall-Gurney field in Kansas, seeking to demonstrate that CO_2 EOR can provide energy, economic and environmental benefits. A companion project underway in the Hall-Gurney field involves testing the feasibility of 4D high resolution seismic monitoring of CO_2 injection in thin, relatively shallow mature carbonate reservoirs. Incorporating this kind of time-lapsed monitoring data into CO_2-EOR programmes could dramatically improve the efficiency and economics of using the technology in many mid-continent fields.

Vocabulary-building activities

1 Natural drives

Oil and gas reservoirs have their own natural drives which allow a degree of recovery to take place without any special assistance or enhancements. Read the following paragraphs about three types of oil reservoir drive and label the three graphs below.

Dissolved Gas Drive

When a well is drilled, pressure on the oil in the reservoir can be released and gas can bubble out of the oil. The pressure of the oil forces the oil through the rock into the well. This type of drive reservoir is very inefficient. It has a very rapid decline in both reservoir pressure and oil production rate, as the oil is produced. There is a very rapid increase in the gas/oil ratio near the end of production.

Free Gas Cap Expansion Drive

In some fields, there is a gas cap above the oil which expands in production and puts the oil under pressure. This type of drive shows a moderate fall in reservoir fluid pressure and production as the oil is produced. The gas/oil ratio rises sharply when the oil has nearly all been produced.

Water Drive

Reservoirs with water drive are driven by the expansion of water below or adjacent to the reservoir. As the oil is produced, it is replaced by water. An active water drive exerts almost constant reservoir pressure and oil production throughout the life of the well. Gas/oil remains the same, but eventually, when the reservoir 'goes to water', water production increases dramatically.

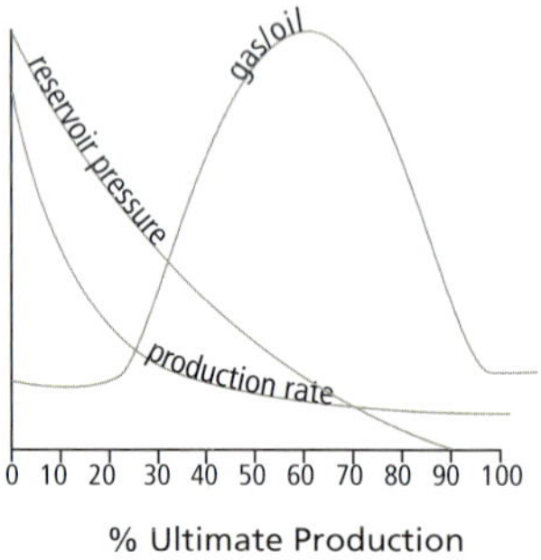

1 ______________________

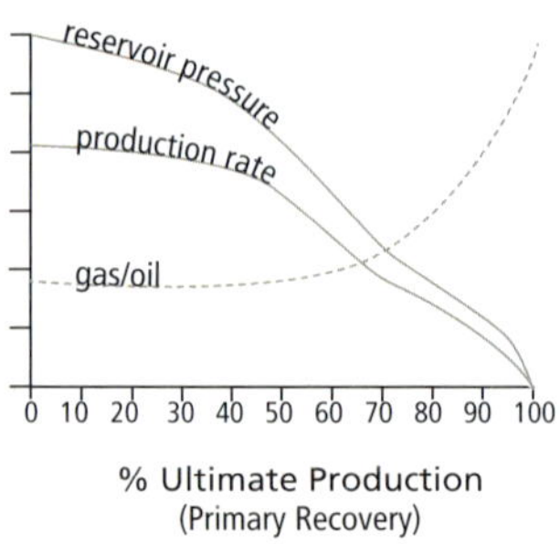

2 ______________________

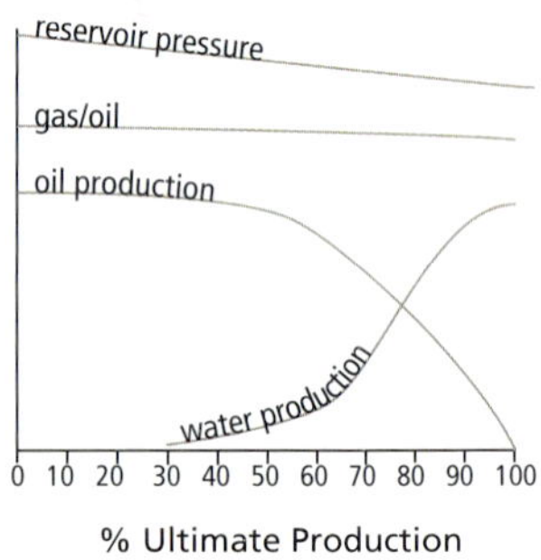

3 ______________________

2 Detailed description

Here is a description of five types of improved recovery. Each has a key phrase missing. Which key phrase belongs to each description and where should it be inserted?

and make it more fluid　in separate slugs　some of the remaining
that dissolves in the oil　to sustain the fire

Once the natural drive of the reservoir is exhausted, oil companies turn to other methods to boost recovery:

1 *Waterflood*: This technique is usually the first to be tried after primary production. It involves injecting water through injection wells. The water forces the oil through to producing wells. It can recover 5 to 50% of the remaining oil in place.

2 *Miscible gas process*: This technique involves injecting a gas (usually carbon dioxide or liquid petroleum gas) into the reservoir. It can recover about 35% of the remaining oil.

3 *Chemical flood*: This technique involves injecting different chemicals, each serving a different purpose, into the depleted reservoir. This recovers around 40% of remaining oil, but can only be used in sandstone reservoirs, not carbonates.

4 *Steamflood*: This technique involves using heat to make heavy oil more fluid for recovery. Steam is injected into the reservoir to heat the oil. Recovery rates can vary from 25% to 60% of the oil in place.

5 *Fire flood*: This is another thermal technique in which subsurface oil is set on fire and air is injected into the reservoir. The oil becomes more fluid and flows towards the producing wells. Approximately 30 to 40% recovery is expected with this method.

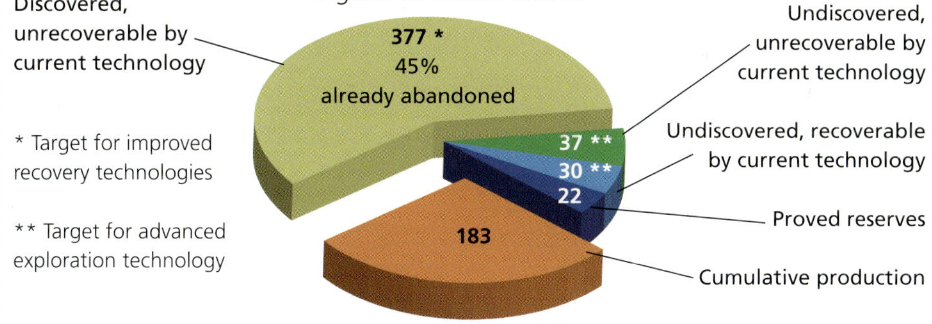

This chart shows how much of the world's production, unrecoverable by current technology, could be recovered using advanced exploration technology.

Source: Cano Petroleum (www.canopetro.com)

Refresh

Refresh the vocabulary you have learnt in this unit.

Put it to work

Describe the recovery methods that are used in any producing reservoirs you know. How do they work? How successful are they?

6 Pipelines

Pipelines are one of the most common ways of transporting crude oil or natural gas to the client, but they carry security and environmental risks and can be the subject of disputes between the countries through which they pass. This unit focuses on vocabulary which will help you to describe pipeline transport in English.

Focused reading

1

Pipeline security

Westminster is a company which specializes in the security of oil and gas pipelines. Here is an extract from their website (www.wi-ltd.com/).
Why is the system 'dual purpose'? How does it work?

Westminster Dual Purpose Pipeline Security and Leak Detection System

The protection of pipelines against sabotage, illegal tapping and terrorist action, combined with the detection of leaks in buried pipelines, etc., is a high priority in all countries, but until now has been notoriously difficult to achieve.

The Dual Purpose Pipeline Security/Leak Detection System is an entirely new advanced early warning system for the detection of third party interference (TPI) tampering and illegal tapping attempts, as well as locating leaks in buried oil, gas and liquid pipelines.

Leaks, damage, illegal tapping and sabotage to pipelines is a worldwide problem costing pipeline operators millions of dollars a year in disruption and lost production, not to mention loss of life that often occurs. Westminster's unique solution to the problem is to provide a Dual Purpose Pipeline Security and Leak Detection System utilizing a single fibre-optic cable run adjacent to the pipeline. This system allows the detection of illegal tapping or sabotage of the pipeline, while also providing the benefit of real-time monitoring against any ruptures or leaks.

Product Benefits

- Early warning of leaks
- Early warning of potential disruption
- Immediate and accurate location of disturbance
- Tamper proof
- Easy to install and low maintenance
- High levels of physical security
- Low risk of service interruptions
- Turnkey system

Source: Westminster International Ltd (www.wi-ltd.com)

2 Frequently asked questions

A sales manager from Westminster, Oliver Matheson, is having a meeting with the head of Pipeline Security from the Aladdin Oil & Gas Company, Vladimir Shukhov. Oliver has introduced the dual purpose system and Vladimir now has some questions for him.

1 Which of the key product benefits listed on the website do they discuss?
2 Are there other benefits, not mentioned on the website, which they also discuss?

VS: Yes, that sounds interesting, Oliver, but is it too good to be true?
OM: Not at all. What are you thinking of Vladimir?
VS: Well, isn't the cable subject to electrical interference?
OM: Not at all. It's completely immune to radiofrequency or electromagnetic interference. It's been tried and tested in many environments.
VS: And how far afield does each control point cover?
OM: Around 60 kilometres and you can network as many controllers as you want, according to the length of the pipeline.
VS: And how accurate are its detection devices?
OM: Well, for breaches of security, we say that it can detect an intrusion to within 50 metres, while for leaks it's as little as one metre.
VS: Regardless of how long the pipeline is?
OM: Exactly. Irrespective of the length of the line.
VS: And you say that it can give early warnings of threats?
OM: Yes, as I said, any intrusion can be spotted up to 60 kilometres away.
VS: What about maintenance?
OM: Well, it's as low-maintenance as it possibly can be. It's extremely robust and highly reliable. And it works equally well in all weather conditions and in all soil types.
VS: OK, so can you give me an idea of the price ...?

3 Pipeline research

Use an online encyclopedia, such as Wikipedia, to research the following pipelines:

- Baku–Novorossiysk
- Brent System
- Kazakhstan–China
- Trans–Alaska
- Yamal–Europe

Make notes on the information provided, such as:

- the building of the pipeline – the date, the ownership, etc.
- technical information – the length, the capacity, the diameter, the pressure.
- any other information – problems, future developments, etc.

Vocabulary-building activities

1 Pipeline systems

The introduction to pipeline systems below describes the main elements of a pipeline system. Read the introduction and then add a few words (keeping to a limit of five words) to each part of the system on the diagram to describe the purpose of that part. The first has been done for you.

Pipeline networks are composed of several pieces of equipment that operate together to move products from one location to another. The main elements of a pipeline system are shown in the diagram below.

- *Initial Injection Station*: Known also as Supply or Inlet Station, is the beginning of the system, where the product is injected into the line. Storage facilities, pumps or compressors are usually located at these locations.
- *Partial Delivery Stations*: Known also as Intermediate Stations, these facilities allow the pipeline operator to deliver part of the product being transported.
- *Compressor/Pump Stations*: Pumps for liquid pipelines and compressors for gas pipelines, are located along the line to move the product through the pipeline. The location of these stations is defined by the topography of the terrain, the type of product being transported or operational conditions of the network.
- *Block Valve Stations*: These are the first line of protection for pipelines. With these valves, the operator can isolate any segment of the line for maintenance work or isolate a rupture or leak. Block valve stations are usually located every 20 to 30 miles (48 km), depending on the type of pipeline.
- *Final Delivery Station*: Known also as Outlet Stations or Terminals, this is where the product will be distributed to the consumer. It could be a tank terminal for liquid pipelines or a connection to a distribution network for gas pipelines.

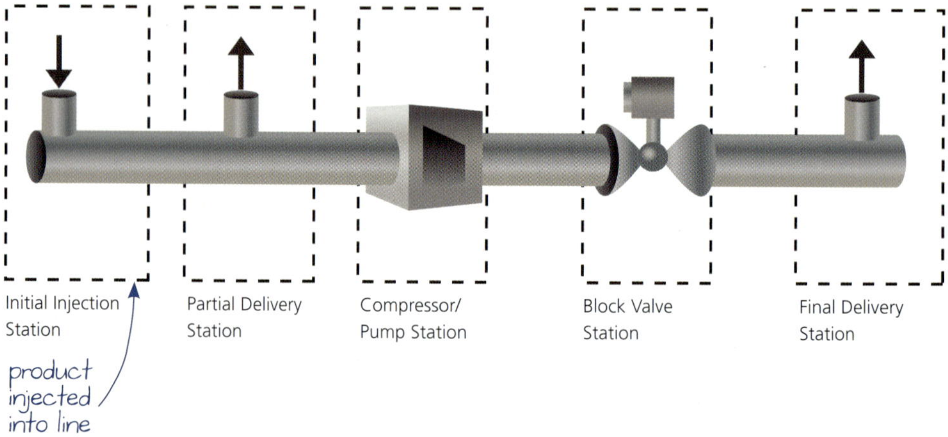

Pipeline schematic

2 Pipeline statistics

Here are some figures about pipelines. Fill in the blanks in the sentences below using the figures given in the box. Use the following approximate conversions to help you.

One mile is around 1.8 kilometres.
There are 12 inches in a foot.
One metre is just under 3 feet.
One inch is approximately 2.5 centimetres.

12 and 24 metres long *2 inches to over 60 inches* *5 and 20 feet per second*
a 6-mile stretch *10 to 120 centimetres* *of about 1 to 2 metres*

1 The first pipeline covered ______________ from an oilfield in Pennsylvania to the nearest railway station.

2 A typical diameter for an oil pipeline ranges from ______________.

3 Pipelines are normally buried at a depth ______________.

4 Oil flows at a speed of between ______________.

5 Natural gas pipelines have a greater range of diameters than oil pipelines, varying from ______________.

6 A typical length of pipeline is between ______________.

3 Word partnerships with adjectives and adverbs

Match the two parts of each sentence. Some may have more than one solution.

1 It is notoriously
2 This is an entirely
3 We are dealing with worldwide
4 We can offer a unique
5 You will have real-time
6 It is completely immune
7 It is an extremely
8 He is a highly
9 It is an absolutely

a to electromagnetic interference.
b system monitoring.
c solution to the problem.
d security problems.
e robust machine.
f reliable employee.
g new system.
h foolproof technique.
i difficult to achieve.

Put it to work

Imagine that you are responsible for security on a stretch of pipeline. Draw a diagram to show a section of your pipeline. On your diagram, include its dimensions and a selection of the equipment used for delivery of the oil or gas. Explain how the pipeline functions. What security measures would you take to protect the pipeline and why?

Refresh

Refresh the vocabulary you have learnt in this unit.

Glossary

Unit 5

Improved oil recovery	
oil recovery	taking oil from a producing well
stranded	oil reservoirs which cannot be recovered or are too expensive to recover
three distinct phases	three different stages of recovery
oil in place	the oil which is still in the reservoir
displace the oil	move the oil from one place to another
easy-to-produce oil	oil which is easy and cheap to recover
commercially successful in varying degrees	more or less commercially successful – sometimes more, sometimes less
thermal recovery	recovery involving the use of heat
to thin the oil	to make the oil less viscous
dissolve in the oil	to become incorporated in the oil to form a solution
long-chained molecules	molecules with many different elements held together to form a chain (also called a polymer)
detergent	a chemical agent which reacts with dirt and impurities to make them more soluble
surfactants	a chemical agent which reduces the surface tension and allows oil to flow more easily
hampered	impeded or slowed down because they are expensive
unpredictability	their effectiveness cannot be forecast or predicted
fertilizer	a chemical product used by farmers to help their crops grow better
time-lapsed	when seismic surveys are carried out with a period of time (often many years) between them, they are called time-lapsed
reservoir drive	the natural pressures in the reservoir which force the oil and gas to move towards the producing wells
bubble out	to come out naturally in bubbles
gas/oil ratio	the ratio between gas and oil in the hydrocarbons that are produced
gas cap	an area of gas trapped in a formation above an area of oil
adjacent to the reservoir	next to the reservoir
goes to water	when oil or gas stops flowing into the producing zone and water flows instead (the well 'goes to water')
slugs	in a chemical flood, different chemicals are injected separately into the well (each separate injection is called a slug)
exhausted	when no more oil or gas can be recovered
to boost recovery	to enhance or improve recovery
miscible	can dissolve to form a homogenous mixture
set on fire	the subsurface oil is burnt (i.e., set on fire) to create heat

Unit 6

Pipeline construction and security	
dual purpose	with two purposes
sabotage	deliberately damaged or destroyed
illegal tapping	stealing (tapping) oil from the pipeline by making a hole in it, which is against the law (illegal)
buried	put under the ground
notoriously difficult to achieve	well known for being difficult to achieve
early warning system	a system which gives a warning of something, some time before it happens
third party interference	when someone who is not the supplier or receiver of the oil in a pipeline tries to take the oil
tampering	interfering with the pipeline to cause damage
locating	identifying the place of something
disruption	interruption or disturbance to the pipeline
fibre-optic cable	a cable containing thin fibres of glass to transmit light signals
ruptures	breaks in the pipeline
tamper proof	cannot be interfered with
turnkey system	when a whole system is bought as a complete system, ready to use
subject to	can suffer from (interference)
immune to interference	does not suffer from interference
how far afield …?	at what distance …?
intrusion	when someone goes near the pipeline who is not allowed to
regardless of …	without considering (how long the pipe is)
irrespective of …	without considering (how long the pipe is)
threats	dangers or risks
spotted	seen, often at a distance
robust	strong and difficult to damage
topography of the terrain	the shape and structure of the land over which the pipeline goes
partial delivery	when some of the oil can be taken out of the pipeline for delivery, but the rest continues to the end of the line
isolate a segment	to cut off a section of pipeline
real-time	taking place as it happens
foolproof	so simple even an idiot (a fool) can use it

7 Tankers

Tankers transport crude oil from their points of production to their points of consumption, which are typically oil refineries, owned and operated by a diverse spread of clients. Worldwide, tankers carry some two billion barrels annually, but amazingly, the cost of tanker transport amounts to only US$0.02 per gallon at the pump. This unit focuses on vocabulary which will help you to describe tanker transport in English.

Focused reading

1 Which media?

Here are extracts from three articles about oil tankers.

1 Which article is from …
- a daily newspaper?
- a specialist oil and gas industry magazine?
- an online encyclopedia?

Extract 1

A lack of spare oil tankers and an increase in OPEC production are driving shipping rates to record highs, said Morten Arntzen, CEO of Overseas Shipholding Group, the largest US-based oil tanker owner. 'This is the first time in 31 years that we've had close to 100% utilization of the world's tanker fleet,' Arntzen said last week. Higher rates caused third-quarter earnings for tanker owners to surge up to ten times over from a year earlier. Overseas Shipholding boosted earnings by a factor of five, and Teekay saw its profit jump more than tenfold. Fourth-quarter earnings may show steeper gains because rates have climbed since late September.

©Stahl-Zentrum

Royal Dutch-Shell Group of Companies chartered the Crown Unity, one of Overseas Shipholding's 22 very large crude carriers, to transport two million barrels of oil from the Arabian Gulf to Europe last week for $183,113 a day, according to data compiled by Bloomberg and a formula on the website of Norway-based ship broker R.S. Platou. That's more than ten times Overseas Shipholding's break-even cost of $17,400 a day for giant crude carriers.

Extract 2

Pakistani port officials warned yesterday that they faced a major oil spill along the southern coastline after a tanker that ran aground in heavy storms began to crack open. A growing slick of oil washed ashore along the main beaches outside Karachi, bringing toxic fumes and hundreds of dead fish, sea birds and turtles. More than 1,000 policemen, equipped with masks, were deployed to close the seafront. Around ten miles of beach, which every evening is normally filled with families, has been closed. The single-hull tanker, Tasman Spirit, ran aground in heavy monsoon storms more than two weeks ago. Salvage experts managed to retrieve some of the 67,500 tonnes of Iranian crude oil it was carrying, but more than 40,000 tonnes remained on board, port officials said. If the rest seeped into the sea, it would become one of the world's worst oil spills.

Copyright Guardian News & Media, 2003, Rory McCarthy

Extract 3

Oil tankers are only one source of oil spills. According to the United States Coast Guard, 35.7% of the volume of oil spilled in the United States from 1991 to 2004 came from tank vessels (ships/barges), 27.6% from facilities and other non-vessels, 19.9% from non-tank vessels, and 9.3% from pipelines; 7.4% came from mystery spills. On the other hand, only 5% of the actual spills came from oil tankers, while 51.8% came from other kinds of vessels. The detailed statistics for 2004 show tank vessels responsible for somewhat less than 5% of the number of total spills, but more than 60% of the volume. In summary, spills are much more rare, but much more serious on tank vessels than on non-tank vessels. The International Tanker Owners Pollution Federation has tracked 9,351 accidental spills that have occurred since 1974. According to this study, most spills result from routine operations, such as loading cargo, discharging cargo and taking on fuel oil. 91% of the operational oil spills are small, resulting in less than seven metric tons per spill. On the other hand, spills resulting from accidents, like collisions, groundings, hull failures and explosions, are much larger, with 84% of these involving losses of over 700 metric tons.

2 If these extracts appeared in a newspaper from your country, which section would each of the articles appear in: National News, International News, Business, Environment or other?

Vocabulary-building activities

1 Third-quarter profits

This is an extract from an informal discussion between two managers from Overseas Shipholding, Rob McAndrew and Barry Delaney. Complete the conversation using information from Focused reading Extract 1 to help you. You will need to change the wording slightly to suit the conversation better, but there are no fixed answers.

RM: Have you seen our profit figures for the third quarter?
BD: Not yet. Is it good news?
RM: Yes, definitely. **(1)** ______________________.
BD: That's excellent.
RM: It is, but it's not as good as Teekay's figures.
BD: Really, how did they do?
RM: **(2)** ______________________.
BD: That's incredible. Why is the industry doing so well?
RM: Well, two reasons really. **(3)** ______________________.
BD: I see. How much do we charge to charter a tanker?
RM: Approximately **(4)** ______________________.
BD: And what's our break-even?
RM: Somewhere in the region of **(5)** ______________________.
BD: So we're making nearly 1,000% profit!
RM: That's about right, yes. Amazing, isn't it!

2 Press conference

Here is part of a transcript of a press conference following the oil spill reported in Focused reading Extract 2. Here, the Pakistan Energy Minister is answering questions from journalists about the spill. Write in the spaces provided what you think his answers should be, using information from the extract.

JOURNALIST: Where exactly did the tanker run aground?
MINISTER: **(1)** ______________________
JOURNALIST: How exactly did it run aground?
MINISTER: **(2)** ______________________
JOURNALIST: Has any oil come ashore yet?
MINISTER: **(3)** ______________________
JOURNALIST: How much oil was on board the tanker?
MINISTER: **(4)** ______________________
JOURNALIST: Did all of it spill into the sea?
MINISTER: **(5)** ______________________
JOURNALIST: Is the beach still open?
MINISTER: **(6)** ______________________
JOURNALIST: Has it caused much environmental damage?
MINISTER: **(7)** ______________________
JOURNALIST: What kind of tanker was the Tasman Spirit?
MINISTER: **(8)** ______________________

JOURNALIST: In many waters, single-hulled tankers have been banned. Why not here in Pakistan?

MINISTER: This is very unfortunate. As you know, we are putting pressure on tanker owners to invest in double-hull tankers, but it is proving to be a difficult process.

3 Sources of oil spills

1 From the information provided in Focused reading Extract 3, complete the following statistical table showing the sources of oil spills in the United States between 1991 and 2004. Write 'not known' where there is insufficient information to complete the table.

Oil Spills in the United States between 1991 and 2004

Source of oil spill	% by volume of oil spilt	% by number of incidents

2 Complete the following key:
Percentage of spills from routine operations, under seven metric tons per spill = ____________ %
Percentage of spills from accidents, over 700 metric tons per spill = ____________ %

3 What can be concluded from the data?

Refresh

Refresh the vocabulary you have learnt in this unit.

Put it to work

There is a proposal to ban oil tankers in the sea near your country. Choose whether you would support or oppose this proposal and prepare notes for a short presentation to make your case. Use slide headings and bullet points to summarize your argument.

Section 5
Producing and Delivering

8 Science in the Oil and Gas Industry

There are many jobs in the oil and gas industry where a scientific background is required. One example is assaying. When a customer purchases crude oil, it is important that they know exactly what they are buying. Assaying is a scientific method that provides benchmarks against which any crude can be compared. This unit focuses on vocabulary which will help you to describe assaying and other scientific work in English.

Focused reading

1 Assaying services

The following extract is from the website of an assaying company called Intertek (www.intertek-cb.com). What services does the company provide?

Top Sites
http://www.intertek-cb.com

Crude Oil Quality Assay Testing Laboratories, Detailed Hydrocarbon Analysis

Full petroleum assay testing services are provided by Intertek through a global network of experienced crude oil laboratories. The Intertek crude oil labs perform detailed hydrocarbon analyses, staffed by expert petroleum chemists with state-of-the-art instrumentation. Intertek is the world's premier independent petroleum laboratory network, with the expertise and capabilities required for full crude oil assay assessments and detailed hydrocarbons analysis. Assay assessment includes:

- Full petroleum assay testing and analysis
- Full assay reporting
- Full crude oil characterization
- Detailed hydrocarbon analysis
- Methanol in crude monitoring
- Mercury in crude analysis
- Enter, smooth and graphically display laboratory crude assay data
- Store published crude assays in a secure database
- Compare graphically current crude assay to reference assay

'Superb support from Intertek allowed crude oil samples to be analyzed and approved in 24 hours while the cargo was still at sea, enabling the crude oil to be optimally utilized by our refineries and enabled the purchase of a second cargo only four days later.'

2

Premium, light and sweet

The following extract is from Statoil's website (www.statoil.com).

In chemical terms, what do you think is the meaning of the words 'premium', 'light' and 'sweet'? (N.B. 'API' is a measurement of crude oil quality from the American Petroleum Institute.)

Top Sites

http://www.statoil.com

Home *News* *Contact*

Crude Oil Assays

Statoil is the world's third largest exporter of crude oil and the leading producer of North Sea crude. We can offer a variety of crudes from a range of more than twenty. Most of the crudes marketed by Statoil are premium, light, sweet grades.

- **Azeri Light** is a light, low sulphur crude oil.
  Region: Asia
  Quality: API between 30–40
- **Dalia** is a medium density, low sulphur crude oil.
  Region: Africa
  Quality: API between 20–30
- **Ekofisk Blend** is a light, low sulphur North Sea crude oil.
  Region: Europe
  Quality: API between 30–40
- **Grane** is a heavy, high TAN North Sea crude oil.
  Region: Europe
  Quality: API below 20
- **Hungo Blend** is a medium density, medium sulphur crude oil.
  Region: Africa
  Quality: API between 20–30
- **Lufeng** is a medium density, low sulphur, waxy crude oil.
  Region: Asia
  Quality: API between 30–40

Source: Statoil (www.statoil.com)

Vocabulary-building activities

1 Fill in the blanks

Complete the passage below using the words in the box.

content	dependent	determined	form	liquefied	proportions
subjected	types	whereas	yield		

Crude oil and natural gas consist of hydrogen and carbon in different **(1)** __________. For example, crude oil contains between 84 and 87% carbon, **(2)** __________ natural gas contains between 65 and 80% carbon.

In crude oil, there are four **(3)** __________ of hydrocarbon molecule:

- the paraffin or alkane molecule
- the napthene or cycloparaffin molecule
- the aromatic or benzene molecule
- the asphaltic molecule

The physical properties of crude oil are **(4)** __________ by the relative percentage of these different hydrocarbon molecules. In turn, the quality and therefore the value of the product are also **(5)** __________ on what can be refined from the crude oil.

Broadly speaking, the greater the **(6)** __________ of asphaltic molecules there is, the more the crude oil can be refined into high-grade gasoline or asphalt. On the other hand, the greater the content of paraffin molecules, the waxier the crude, and therefore the greater the **(7)** __________ of lubricating oil and kerosene.

In refining, oil is **(8)** __________ to heat in a distilling column and products are extracted in liquid **(9)** __________ as they cool, with different products being extracted at different temperatures. The most demanded product in refining is gasoline, followed by heating fuel, jet fuel and then a number of less-produced products, including **(10)** __________ gases and various petro-chemical feedstocks, such as methane, ethylene and propylene.

2 Scientific properties

Reread Focused reading 2, from Statoil's website, and then read the extract opposite about how oil and gas scientists define the properties of crude oil. Answer the questions that follow.

When purchasing crude oil, refiners are interested in the following information:

Origin. The oilfield of origin indicates the quality of the crude. Oil companies often sell a blend of crudes from different fields.

API. This stands for American Petroleum Institute and is a measure of the density of the oil. The higher the number, the lighter the oil. Light oils, often transparent in colour and very fluid, have an API of between 35 and 45. Heavy oils, which are very viscous and dark-coloured have an API of below 25.

Sulphur content. Sulphur is an undesirable element in crude oil and needs to be removed during the refining process. Crude oil is called sweet or sour, depending on its sulphur content. Sweet crudes have less than 1% sulphur content by weight, sour crudes somewhere over 1%.

Pour point. This indicates the amount of wax in a crude oil and is the lowest point at which the oil will still pour before it solidifies. A waxy crude oil is found in liquid form below the surface before it is extracted, but can solidify as it is brought out of the ground and this can lead to clogging in the tubing of the well.

Crude oils are often measured against a benchmark, such as Brent from the North Sea, which has an API of 33 and a sulphur content of 0.3%.

1 Are there any blended crudes on the list?
2 Which is the heaviest oil on the list?
3 Which oils would be transparent in colour?
4 Which is the most sour crude on the list?
5 Which crude might clog as it is extracted from the ground?
6 Which crudes are closest to the Brent crude benchmark?

3 Assaying

Assaying is one of hundreds of different functions carried out during the exploration and production of oil and gas. These operations are all involved in different kinds of testing or analysis. Which do you do in your daily work? In what situations? Do you do any other scientific operations not mentioned here?

- analyzing
- gauging
- monitoring
- researching
- assaying
- interpreting
- plotting
- sampling
- benchmarking
- logging
- recording
- surveying
- disseminating
- measuring
- reporting

Put it to work

You have been asked to prepare a presentation for a group of science graduates interested in working in the oil and gas industry. Make notes for your presentation about how a degree in science can be helpful in the oil and gas industry.

Refresh

Refresh the vocabulary you have learnt in this unit.

Glossary

Unit 7

Transportation by tanker	
tankers	ships used for transporting oil
crude oil	oil before it has been refined (or separated into different products)
oil refineries	places where oil is refined
lack of spare tankers	not enough tankers available for use
shipping rates	the cost of transportation by ship
tanker fleet	a group of tankers under the same ownership
chartered	hired to carry oil from one place to another
single-hull tanker	a tanker with one hull – most tankers nowadays are double-hull to reduce the risk of pollution in the event of a collision
tank vessels	another word for oil tankers
tracked	recorded
cargo	the goods on a ship, train or plane (on an oil tanker, the cargo is oil)
discharging	unloading the oil from the tanker
come ashore	to move from the sea to the land
Tanker accidents	
ran aground	when the sea is too shallow for a ship, it runs aground (i.e., it gets stuck on the seabed)
slick of oil	a volume of oil which has escaped from a ship into the sea
toxic fumes	gas, smoke or vapour which smells bad or is dangerous to inhale
salvage experts	people who retrieve ships which have run aground or sunk and the goods on the ships
seeped	escaped (into the sea) slowly
mystery spills	oil spills with no clear explanation
collision	when two tankers hit each other
groundings	incidents when tankers run aground
Economies of scale, pricing and profits	
record highs	peaks of profits which have never been reached before
third-quarter earnings	profits made in the third quarter of the financial year
surge	a big increase
boosted earnings	improved profits
a factor of five	five times more
jump tenfold	increase by ten times
steeper gains	bigger increases
break-even cost	the amount of money that has to be made to cover investment costs

Unit 8

Chemical analysis of oil	
assaying	carrying out chemical tests on crude oil (or other products) to check its characteristics against a standard
perform analyses	to carry out tests and interpret the data (singular = 'analysis', plural = 'analyses')
characterization	the analysis of crude oil to show its different characteristics
to smooth data	to make the data easier to understand
current crude assay	the characteristics of the crude oil under examination
reference crude assay	the characteristics of a standard or benchmark crude oil
optimally utilized	used as well as possible
sweet	low in sulphur
liquefied	turned into liquid form
yield	the quantity produced
waxier	with a higher content of wax
distilling column	the central part of a refinery where oil is distilled into its component parts
feedstocks	chemicals refined from hydrocarbons and then used to produce different kinds of petro-chemicals
blend	mixture (of different crude oils)
transparent	when a substance allows light through so that you can see objects behind it
undesirable	not wanted
sour	the opposite of sweet; with a high sulphur content
pour point	the lowest temperature at which a crude oil will still flow
solidifies	becomes solid (the opposite of liquefies)
benchmark	the standard by which other crude oils are measured

Word List

Word	Section/Unit
crest *(n)*	4.2
criticality *(n)*	2.6
cross section *(n)*	4.4
crown block *(n)*	3.1
crucial *(adj)*	1.6
crude *(adj)*	5.7
cumulative *(adj)*	4.3
current *(n)*	2.2
cuttings *(n)*	3.1
cylinder *(n)*	3.8
D	
dead line *(n)*	3.2
debris *(n)*	2.7
decibel *(n)*	3.8
deck *(n)*	2.2
decommission *(v)*	2.7
defect *(n)*	2.6
deform *(v)*	4.2
degree *(n)*	5.5
delegate *(n* or *v)*	2.3
demand *(n* or *v)*	1.1
demanding *(adj)*	2.3
dense *(adj)*	4.7
density *(n)*	4.7
depletion *(n)*	4.3
deposit *(n* or *v)*	4.1
deposition *(n)*	4.2
depreciation *(n)*	1.4
derrick *(n)*	2.2
detach *(v)*	2.7
detection *(n)*	2.5
detergent *(n)*	3.5
deteriorate *(v)*	2.4
determine *(v)*	2.3
detonation *(n)*	5.1
deviation *(n)*	3.6
diameter *(n)*	3.1
diffusion *(n)*	5.2
digital *(adj)*	4.4
dip *(n* or *v)*	4.2
directional *(adj)*	3.8
dislodge *(v)*	2.5
disperse *(v)*	3.5
displace *(v)*	3.8
disruption *(n)*	5.6
disruptive *(adj)*	3.8
dissolve *(v)*	5.5
distill *(v)*	5.8
distinct *(adj)*	5.5
dogleg *(n)*	3.6
downstream *(n* or *adj)*	1.1
downtime *(n)*	3.3
drawworks *(n)*	3.2
drill collars *(n)*	3.1
drillstring *(n)*	3.1
drilling line *(n)*	3.1
droplet *(n)*	5.2
drum *(n)*	3.2
dual purpose *(adj)*	5.6
E	
echo *(n* or *v)*	4.5
eddy *(n)*	3.3
effectiveness *(n)*	1.2
efficiency *(n)*	1.6
efficient *(adj)*	1.7
elapse *(v)*	1.5
embrittlement *(n)*	3.6
emission *(n)*	1.6
emit *(v)*	4.5
emulsion *(n)*	3.5
encase *(v)*	4.2
encounter *(n* or *v)*	4.7
endowment *(n)*	4.3
epoxy *(n, adj* or *v)*	5.3
equipment *(n)*	1.4
etch *(v)*	5.4
evaporites *(n)*	4.1
evaporitic *(adj)*	4.8
exhaust *(v)*	5.5
expandable *(adj)*	4.7
expansion *(n)*	1.2
expel *(v)*	4.1
expenditure *(n)*	1.4
expertise *(n)*	1.2
expire *(v)*	1.5
explore *(v)*	1.1
exposure *(n)*	2.5
extinguish *(v)*	2.5
extract *(v)*	1.1
extrapolation *(n)*	4.6
F	
failure *(n)*	2.6
farm-in *(v)*	4.3
fasten *(v)*	3.2
fast line *(n)*	3.2
fault *(n)*	4.2
feedstock *(n)*	5.8
fertilize *(v)*	5.5
fibre-optic *(adj)*	5.6
film *(n)*	3.5
fish *(n)*	3.6
flank *(n)*	4.2
flare stack *(n)*	2.2
flaring *(n)*	1.6
flaw *(n)*	2.6
fleet *(n)*	5.7
fluctuation *(n)*	1.4
foam-free *(adj)*	5.2
foolproof *(adj)*	5.6
forefront *(n)*	3.3
formation *(n)*	4.7
frac pack *(n)*	5.1
fractional *(adj)*	5.2
fracture *(n* or *v)*	4.1
fuse *(n)*	3.3
G	
gamma ray *(n)*	4.7
gauge *(v)*	5.2
gauger *(n)*	5.2
geologic *(adj)*	4.4
geophone *(n)*	4.5
geosteering *(n)*	4.7
gravel *(n)*	5.1
gravity *(n)*	5.2
greenhouse gas *(n)*	1.6
groove *(n)*	3.6
grounding *(n)*	5.7
guide shoe *(n)*	5.1
guy line *(n)*	5.3
H	
halt *(v)*	4.2
hamper *(v)*	5.5
handle *(n* or *v)*	2.3
harness *(n* or *v)*	2.6

Word	Section/Unit
hazard *(n)*	2.4
heli-deck *(n)*	2.2
helmet *(n)*	2.5
heterogenous *(adj)*	5.4
highly *(adv)*	2.3
hitch *(n)*	2.4
hoist *(v)*	3.2
hoisting line *(n)*	3.1
holiday *(n)*	5.3
hook *(n)*	3.1
hook up *(v)*	2.7
hose *(n)*	3.1
hull *(n)*	2.2
hydrate *(v)*	3.5
hydraulic *(adj)*	3.7
hydrocarbon *(n)*	4.1
hydrophone *(n)*	4.5
hydrostatic *(adj)*	5.1
I	
ignite *(v)*	2.5
illegal *(adj)*	5.6
immune *(v)*	5.6
impair *(v)*	2.4
impedance *(n)*	4.5
impede *(v)*	4.1
impenetrable *(adj)*	4.2
impinge *(v)*	4.8
implementation *(n)*	1.3
improvement *(n)*	3.3
impulse *(v)*	4.5
impurity *(n)*	5.2
in force *(adj)*	1.5
induce *(v)*	4.7
induction *(n)*	3.3
inhibit *(v)*	3.6
innovation *(n)*	3.3
innovative *(adj)*	1.7
installation *(n)*	2.7
instrumentation *(n)*	4.6
integration *(n)*	1.1
integrity *(n)*	5.1
interface *(n)*	2.8
interference *(n)*	5.6
intersect *(v)*	5.4
intrusion *(n)*	1.6
invade *(v)*	4.2
invent *(v)*	1.7
irrespective *(adj)*	5.6
isolate *(v)*	5.6
isopach *(adj* or *n)*	4.4
J	
jacket *(n)*	2.2
jar *(n* or *v)*	3.6
joint venture *(n)*	1.1
K	
kelly *(n)*	3.1
kelly bushing *(n)*	3.4
kerogen *(n)*	4.1
kick *(n* or *v)*	3.6
kill *(v)*	3.6
knocked *(v)*	2.5
knockout *(v)*	5.2
L	
lack *(n* or *v)*	2.4
lagged *(adj)*	4.6
latch *(v)*	3.6
lease *(n)*	1.5
lever *(n)*	3.2
liability *(n)*	1.4
license *(n)*	4.3
limestone *(n)*	4.1
liquify *(v)*	5.8
lithology *(n)*	4.6
loan *(n* or *v)*	1.8
lubricate *(v)*	3.5
lubricity *(n)*	3.5
M	
machinery *(v)*	1.4
magnetometer *(n)*	4.7
maintenance *(n)*	2.6
management *(n)*	1.6
manipulate *(v)*	2.8
manoeuvre *(v)*	3.2
mast *(n)*	3.2
master bushing *(n)*	3.4
maturation *(n)*	4.1
middleman *(n)*	1.3
migration *(n)*	4.1
milestone *(n)*	1.8
mist *(n)*	5.2
model *(n)*	4.4
modernize *(v)*	3.3
module *(n)*	2.2
molecule *(n)*	5.5
motor *(n)*	3.3
mount *(n* or *v)*	2.2
mud logging *(v)*	4.6
multilateral *(adj)*	3.8
N	
negotiate *(v)*	1.5
net *(adj)*	4.4
neutron *(n)*	4.7
node *(n)*	3.2
nozzle *(n)*	3.5
O	
offload *(v)*	2.2
offset *(v)*	4.5
offshoot *(n)*	3.8
offshore *(n* or *adj)*	1.1
on-stream *(adj)*	1.1
onshore *(n* or *adj)*	1.1
operations *(n)*	1.1
optimal *(adj)*	5.8
orientation *(n)*	4.4
orifice *(n)*	5.2
outrigger *(n)*	5.3
overhaul *(v)*	2.6
overhead *(n)*	1.4
overshoot *(v)*	3.6
P	
package *(n)*	1.8
packer *(n)*	5.3
paraffin *(n)*	5.3
pay zone *(v)*	4.4
peak *(n* or *v)*	1.7
penetration *(n)*	4.6
perforate *(v)*	5.1
performance *(n)*	1.2
permafrost *(n)*	2.1
permeability *(n)*	4.1

Word	Section/Unit
phase *(n)*	5.5
piles *(n)*	2.2
pinch-out *(n)*	4.2
pioneering *(adj)*	3.3
pipe seal ram *(n)*	3.7
piston *(n)*	5.3
pit *(n)*	3.5
pixler *(adj)*	4.6
plot *(v)*	4.6
plug *(n or v)*	5.1
porosity *(n)*	4.1
pose *(n or v)*	2.5
potential *(n or adj)*	4.3
pour *(v)*	5.8
premises *(n)*	1.5
pressure *(n)*	3.1
prime mover *(n)*	3.1
procedure *(n)*	2.5
producer *(n)*	1.1
proficiency *(n)*	2.5
proof *(n)*	5.6
property *(n)*	1.4
proppant *(n)*	5.1
prospect *(n or v)*	4.3
protect *(v)*	1.6
pulley *(n)*	3.2
pulse *(n)*	4.7
purge *(v)*	3.3
R	
radial *(adj)*	5.4
ram *(n)*	3.7
ratio *(n)*	4.6
reactivate *(v)*	5.4
real-time *(n or adj)*	4.7
reamer *(n)*	3.6
receiver *(n)*	4.5
recognition *(n)*	1.2
recognize *(v)*	2.5
recycle *(v)*	1.6
redundancy *(n)*	2.8
reef *(n)*	4.1
reel *(n)*	3.2
refinery *(n)*	5.7
reflect *(v)*	4.5
reflection *(n)*	4.5
register *(v)*	1.2
reinforce *(v)*	2.2
relay *(v)*	2.3
remote *(adj)*	2.1
rename *(v)*	1.7
renewal *(n)*	1.2
replace *(v)*	2.6
replenish *(v)*	2.6
requirement *(n)*	1.8
reserve *(n)*	1.1
reservoir *(n)*	4.1
resin *(n)*	5.3
resistivity *(n)*	4.7
respirator *(n)*	2.5
retard *(v)*	5.1
retrieve *(v)*	3.6
revamp *(v)*	3.3
ridge *(n)*	4.1
rift *(n)*	4.2
rise *(v)*	1.7
riser *(n)*	2.8
robust *(adj)*	5.6
roller *(n)*	5.3
rotary hose *(n)*	3.5
rotary table *(n)*	3.1
rubble *(n)*	3.8
rupture *(n or v)*	2.5
S	
sabotage *(n or v)*	5.6
saddle *(n or v)*	4.2
salinity *(n)*	4.8
salvage *(n or v)*	5.7
sandstone *(n)*	4.1
saturate *(v)*	4.1
scale *(n)*	4.4
scintillation *(n)*	4.7
scrap *(n or v)*	2.7
scratcher *(n or v)*	5.1
screw *(v)*	3.4
seabed *(n)*	2.7
seal *(n or v)*	3.5
sector *(n)*	2.1
sedimentary *(adj)*	4.1
seep *(v)*	5.7
segment *(n)*	5.6
seismic *(adj or n)*	4.4
seize *(v)*	3.5
sensor *(n)*	2.8
separator *(n)*	5.2
set up *(v)*	2.3
setting *(n)*	4.3
settle *(v)*	5.2
settling pit *(n)*	3.5
shaft *(n)*	3.2
shake *(v)*	5.2
shale *(n)*	3.5
shale shaker *(n)*	3.5
sheaf/sheaves *(n)*	3.2
shear *(v or adj)*	3.7
shear ram *(n)*	3.7
shield *(n)*	2.5
shoring *(n)*	3.4
shoreline *(n)*	4.1
shutdown *(n or v)*	2.6
shut in *(n or v)*	3.7
sidetrack *(n or v)*	5.3
sidewall *(n)*	4.6
signal *(n)*	4.5
sink *(v)*	2.2
sintered *(adj)*	5.4
slant *(n or v)*	3.8
slick *(n)*	5.7
slimhole *(n)*	3.8
slimy *(adj)*	5.4
slip *(n or v)*	4.2
slip ram *(n)*	3.7
slotted liner *(n)*	5.1
slough *(v)*	3.6
slug *(n)*	5.5
slurry *(n)*	5.1
solidify *(v)*	5.8
solubility *(n)*	3.5
sonde *(n)*	4.6
source *(n)*	4.1
spare part *(n)*	2.6
spear *(n)*	3.6
spill *(n or v)*	1.6
spin *(v)*	5.2
spin-off *(n or v)*	1.7
spiral *(n)*	3.6
split *(n or v)*	5.3

Answer Key

Section 1

Unit 1

Focused reading

1. exploration and production (E&P), mergers and acquisitions, supply and demand
2. **a** 2, 4 **b** 1, 3 **c** 5
3. In October 2008, all of these statements were true. Sentences 5 and 6 are historic, so will be correct.

Vocabulary-building activities

1. 1 Demand for oil will continue to increase until 2015.
   2 The market for liquefied natural gas is growing steadily.
   3 Consumption of fossil fuels has declined slightly in recent years.
   4 The supply of North Sea Oil can be maintained until around 2024.
   5 Transportation of oil presents many security risks.
   6 Reserves of natural gas are being found almost daily.
   7 Drilling for oil and gas is a major part of our upstream operations.
   8 The extraction of oil at the Burgan field began in 1946.
   9 Exploration for oil and gas has been going on for over a hundred years.
   10 Production of gas from deep-water reserves is our main area of expertise.

2.

Verb	Noun (abstract)	Noun (person)
operate	operation	operator
extract	**extraction**	
compete	competition	competitor
acquire	acquisition	
consume	**consumption**	consumer
produce	**production**	**producer**
explore	**exploration**	**explorer**
subcontract		**subcontractor**
supply	supply	**supplier**
transport	**transportation**	

3. 1 located
   2 groups
   3 exploration
   4 ownership
   5 refineries
   6 market
   7 consumers

Unit 2

Focused reading

1. Answers might include:
   wide recognition, firm position, tailor-made services, major, grown consistently, success, commitment, highest European standards, entire, continuous growth and stabilization, recognition in Europe, presence at worldwide events, cutting-edge technology
2. Students' own answers.
3. 1 The aim of the presentation is to give a positive impression of the company and its performance.
   2 The audience could be shareholders or potential investors in the company, for example.

Vocabulary-building activities

1. 1 1 multinational
     2 market leader
     3 rapid improvements
     4 contract out
     5 joint ventures
   2 marketing strategy, history and track record, size, international profile
2. a wide range of services
   commitment to excellence
   environmentally sound
   founded in 1979
   grouped into four geographical areas
   key competitive advantages
   leading-edge technology
   proven track record
   vital to our success
3. 1 founded in 1979
   2 grouped into four geographical areas
   3 proven track record
   4 a wide range of services
   5 key competitive advantages
   6 environmentally sound

7 commitment to excellence
8 vital to our success

4. Model answer:
In our upstream division, we had oil production growth of 6.4%, well above the industry average of 2.4%. We replaced 137% of our production with new oil reserves and nine new licences were acquired as part of our strategy for long-term renewal. In the downstream area, our refining throughput increased by 10% and we completed two major refinery upgrades. There was a focused expansion of our retail network. As far as our portfolio management is concerned, the assets of three of our companies have now been sold. We have had strong revenues and a good profit performance, leading to a significant dividend recommended for shareholders. We are undertaking a corporate restructuring.

Unit 3

Focused reading

1. 1 place
2 price
3 product
4 promotion

2. Promotion – the job involves advertising and public relations.

3. Price – Brian is asking for Keith's advice on the pricing of a proposal he is working on. Keith works in the marketing department.

Vocabulary-building activities

1. 1 promotion
2 price
3 product
4 place or promotion
5 product
6 promotion
7 place

2. Students' own answers.

3. 1 up our prices
2 giving them a good price
3 set the price
4 price sensitive
5 pricey
6 price them out of the market
7 price-fixing
8 agree a price

4. 1 productivity
2 productive
3 production
4 produce
5 producing
6 products
7 producers

Unit 4

Focused reading

1. The balance sheet.

2. *'Successful efforts' accounting*
Balance sheet:
- property acquisitions
- successful exploratory wells
- development costs (including development dry holes)
- support equipment and facilities

P&L:
- unsuccessful exploratory wells and items such as overheads and direct costs

'Full-cost accounting'
Balance sheet:
- all operations, whether successful or unsuccessful

P&L:
- overheads and direct costs of E&P

The treatment of items such as depreciation, depletion of any unproved reserves, or the costs of clean-up operations, varies widely from company to company.

3. £525,000 (the sum of the first two items on the budget)

Vocabulary-building activities

1. 1 incur unnecessary costs
2 unsuccessful exploratory wells
3 charged to expense
4 depletion of unproved reserves
5 cross-border mergers and acquisitions
6 highly sophisticated operation

2. 1 accountant
   2 account
   3 accounts / accounting
   4 accountancy
   5 accounts
   6 account for
   7 accountancy / accounting
   8 accountable to
3. 1 False. The first three items amount to £945,000, much more than half of the initial fixed investments.
   2 True. Contingencies of 10% have been included.
   3 False. Equipment (£25,000) is a small proportion of the budget.
   4 True. £50,000 has been included for working capital.
   5 False. The total is around 90%.

4.

£ million	
450,000	four hundred and fifty thousand
75,000	seventy-five thousand
420,000	four hundred and twenty thousand
25,000	twenty-five thousand
70,000	seventy thousand
96,000	ninety-six thousand
113,600	one hundred and thirteen thousand, six hundred
1,249,600	one million, two hundred and forty-nine thousand, six hundred
16,500	sixteen thousand, five hundred
50,000	fifty thousand
1,316,100	one million, three hundred and sixteen thousand, one hundred

Unit 5

Focused reading

1. The landowner and the oil and gas E&P company.
2. 1 Phrases or extracts to highlight:
   '... in force for a term of FIVE (5) YEARS ...'
   '... not more than One Hundred Twenty (120) days shall elapse between the completion or abandonment of one well and the beginning of operations for the drilling of a subsequent well.'
   '... this lease shall not terminate if Lessee commences additional drilling or reworking operations within One Hundred Twenty (120) days from date of cessation of production ...'
   2 My proposal would be ...
   What about ...
   Let's call it ...
   How does that sound?
   Is that what you have in mind?
   Let's ...

Vocabulary-building activities

1. Paragraph 2 – the company can continue working on the site even if there has been no production.
2. 1 There are four other words beginning with 'there' or 'there':
   Thereafter – after 'the term of five years from the date of this lease'
   Hereinafter – in this contract after this clause
   Thereon – on 'the leased premises or on acreage pooled therewith'
   Thereof – of 'oil and gas on said land or on acreage pooled therewith'
   2 Other typical legal words:
   It is agreed that ...
   remain in force
   of whatsoever nature or kind
   said leased premises
   the expiration of the primary term
   Lessee
   continuously prosecuted
   shall elapse
   said land
   If the production thereof should cease ...
   date of cessation
   If oil or gas shall be discovered ...
   such operations
3. 7, 6, 1, 8, 5, 2, 4, 3

4.

Contract	The drilling area	E&P operations
lease	premises	discovered
expiration	acreage	produced
term	pooled	abandonment
in force	land	completion
terminate		cessation of production

5. 1 terminate
2 premises
3 cessation of production
4 in force, abandonment
5 term

Unit 6

Focused reading

1. Oil spills, damaged land and incidents of air and water pollution.

2. Possible words:
… by using energy efficiently / carefully. (adverb)
… poor / bad air quality … (adjective)
… topical environmental issues, especially / particularly for large population centres. (adverb)
… the provision of clean / better transport fuel … (adjective)
… areas are government / trust-protected … (adverb, adjective)
… to be of great / significant biodiversity value. (adjective)
… the most useful / practical ways … (adjective)
… exercise effective / careful / long-term stewardship of the environment … (adjective)
… delivering real / long-term commercial business improvement … (adjective)

Vocabulary-building activities

1. Possible answers from the following:
Onshore drilling: noise, soil contamination, disturbance to wildlife, erosion, damage to habitat, greenhouse gases, leaks, visual intrusion
Offshore drilling: water pollution, greenhouse gases, spillages, disturbance to wildlife

2. 1 disturbance to wildlife
2 spillages
3 noise, visual intrusion
4 greenhouse gases
5 erosion

3. … by using energy efficiently.
… poor air quality …
… topical environmental issues, particularly for large population centres.
… the provision of cleaner transport fuel …
… areas are environmentally sensitive …
… high biodiversity value.
… the most basic ways …
… exercise responsible stewardship of the environment …
… delivering continuous commercial business improvement …

4. We operate in environmentally sensitive areas of the desert.
We must reduce carbon emissions caused by flaring.
It is important to avoid spillages by handling toxic materials with care.
We measure energy efficiency in joules per annum.
We aim for continuous improvement to our environmental policies.

5. 1 has invested
2 can be pursued
3 has been made
4 are being used
5 has been expanded
6 have been drawn

Unit 7

Focused reading

1. Possible answers: efficient, ruthless, powerful, innovative, clever, wealthy, competitive, lucky

2. *Mergers:*
- Socono merged with Vacuum to form Mobile.
- The independent company Richfield merged to form Atlantic Richfield (ARCO).

Spin-off/Acquisition:
- Atlantic Richfield operations were spun off and bought by Sunoco.

Acquisition:

- Kyso was acquired by Standard Oil of California.

Vocabulary-building activities

1. 1 1954
2 1859
3 1931
4 1989
5 1903
6 1870
7 1973
8 1938
9 1895
10 1997
11 1879

2. 1 e
2 b
3 f
4 d
5 not illustrated
6 a
7 not illustrated
8 c

Unit 8

Focused reading

1. 1 a MOL must comply with environmental conditions within three years.
b They are borrowing this amount from EBRD.
c This is the total amount they have borrowed from all sources.
d This is the amount they have borrowed from commercial banks.
e This is the amount they intend to invest to reach compliance with the bank's conditions.

2 Michel-Marc Delcommune is the CFO of MOL and Alain Pilloux is the business group director for the EBRD. Although the text doesn't make it completely clear, they were probably involved together in the negotiation of the terms and conditions of the agreement.

Vocabulary-building activities

1. 1 True
2 False. The company has to reach compliance with all of the 'applicable norms', under the terms of its entry into the EU.
3 False. MOL expects to reach compliance within three years.
4 False. EBRD requires MOL to adhere to the Environmental Action Plan.
5 False. MOL has agreed to carry out adequate due diligence.
6 False. The banks would like to encourage MOL to adhere to environmental norms.
7 False. MOL will maintain quality leadership and meet high standards.

2.

Bank	Oil and gas company	Both parties
lend money	borrow money	sign a loan agreement
provide financing	obtain a financing package	negotiate a loan
offer funding	raise money	
give a loan	secure a loan	
set interest rates	get financing	

3. 1 achieve, meet, reach
2 achieve, fulfil, meet
3 achieve, ensure, maintain
4 adhere to, comply with
5 achieve, meet, satisfy
6 achieve, adhere to, comply with, maintain
7 fulfil, satisfy

4. as long as, assuming, if, provided that, providing, when

Section 2

Unit 1

Focused reading

1/2. Students' own answers.

Vocabulary-building activities

1. 1 a located at
b off the coast of
c in the Norwegian sector of
d from the northern edge of
e along the north-east coast of
f in northern
g located between

2 a at
b off, of
c in, of
d from, of
e along, of
f in
g between

2. Offshore – driving winds, gale-force winds, hurricanes, lashing rain
Desert – dry and arid, sandstorms
Jungle – monsoons, hot and humid, tropical cyclone
Arctic – blizzards, freezing cold, permafrost, ice-floes

3. 1 jungle
2 arctic
3 desert
4 offshore
5 arctic
6 desert
7 offshore
8 jungle

Unit 2

Focused reading

1. 1 drilling barge, jack-up rig, semi-submersible rig, drillship, compliant platform
2 gravity-base fixed production platform, steel-jacket platform, tension-leg platform
3 drilling barge, gravity-base fixed production platform, steel-jacket platform, tension-leg platform
4 jack-up rig, semi-submersible rig, drillship, compliant platform

2. 1 jack-up rig
2 tension-leg platform
3 semi-submersible rig
4 drillship
5 drilling barge

3. To store oil and then transfer it to tankers or pipelines. They are more economical in small fields or where the lifetime of a field is expected to be short.

Vocabulary-building activities

1. The following are visible: accommodation modules, drilling rig, flare stack, service cranes, drilling pipe deck, heli-deck, store areas

2. 1 heli-deck
2 service cranes
3 store areas
4 flare stack
5 ballast tanks
6 derrick
7 jacket
8 piles
9 barge
10 accommodation modules
11 drilling rig
12 drilling pipe deck

3. 1 False. It is a support vessel for a drilling rig.
2 True
3 True
4 False. It is useful in remote or deep-water locations.
5 True
6 True
7 False. It works well in fields which can be exhausted in a few years.
8 True
9 False

4. 1 a takes off
b shallow
c off-duty
d light
e floating
f filled
g vertical
h hollow
i float

2 uneconomical, ineffective, unexpected, inexpensive, unflexible, unnecessary, unsuitable

Unit 3

Focused reading

1. Drillers

- moving, setting up and operating drilling equipment
- move rigs onto site
- monitor equipment
- assist in crane operations
- operates the brake and motor of the drill
- determines how fast and deep the drill needs to go
- record drilling progress
- play a 'troubleshooting' role

Toolpushers

- monitor the drilling operations
- relay their instructions to drillers
- manage personnel and equipment
- set the drilling schedule
- monitors the schedule
- delegating work

2. The field engineer should be able to manage a high degree of responsibility, as he/she must deal with customers and be responsible for critical and expensive operations. He/she has to have a good technical knowledge of the equipment and its layout, installation, etc., and be able to operate it and instruct others in its operation. He/she must be able to assist in resolving unusual problems.

Vocabulary-building activities

1.

Job title	Job description	Job type
company man	represents the oil and gas operator	management
driller	handles the rotary drilling equipment	technical
installation manager	is in charge of the platform	management
mud logger	analyzes samples of rock cuttings	technical
assistant driller	carries out drilling operations under supervision	technical
roustabout	handles supplies and equipment	manual
wireline operator	operates the various electronic tools used in the well	technical
crane operator	sets up and operates lifting equipment	technical
toolpusher	supervises operations on the rig	management
chief engineer	is responsible for overall operation of the drilling rig	management
roughneck	handles pipes and other equipment for the rig	manual

2. Relational: 1, 2, 8
Communication: 5, 6, 7
Adaptability: 3, 4, 9, 10

3. Students' own answers.

Unit 4

Focused reading

1. 1 travel opportunities
2 onshore leave
3 women working on a rig
4 shift work

2. Wherever possible, workers should not work shifts of more than 14 days and they should not do rollover shifts.

Vocabulary-building activities

1. Possible answers:
Positive: 1, 2, 6
Negative: 3, 4, 5, 7, 8, 9, 10

2. Students' own answers.

3. Model answers:
1 can lead to
2 could result in
3 could bring about
4 can cause
5 can expose rig workers to

4. 1 a impair, reduce, affect
b make
c suffer, decline, deteriorate

2 cause, result in
3 Rollover shifts seem to cause/result in a reduction in alertness.

Unit 5

Focused reading

1. 1 b, 2 d, 3 c, 4 a

Vocabulary-building activities

1. 1 face shield
2 safety goggles
3 rubber gloves
4 high-visibility jacket
5 protective footwear
6 earmuffs
7 hard hat
8 respirator mask

2. 1 sources and hazards
2 detection methods
3 warnings
4 symptoms
5 first-aid procedures
6 personal protective equipment

3. 1 A young pipeline operator was knocked unconscious.
2 Another experienced operator was badly injured.
3 The first drilling schedule was completely disrupted.
4 A chemicals storage area was totally destroyed.
5 Some essential piping was slightly damaged.
6 The local fire crew was informed immediately (or immediately informed).
7 Fortunately, the fire was quickly extinguished (or extinguished quickly).

Unit 6

Focused reading

1. Anthony Johnson's approach is very casual and could be inefficient. However, it is also probably very cheap and well-suited to a small operation.

Faisal Riyad's approach is systematic and efficient, but could lead to waste because of unnecessary shutdowns and unnecessary maintenance.

Pramana Batak's approach takes advantage of the latest technology to ensure that only essential maintenance is carried out.

Colin Freyer does not really explain an approach to maintenance. His work is potentially dangerous, but essentially he does the work he is asked to do.

Igor Borovsky also doesn't really explain an approach. His experience is in using equipment which takes away the need for the rope and harness work that Colin does.

The question of who should be chosen for the job is a matter of opinion and will depend on the exact nature of the job. However, Praman Batak and Igor Borovsky generally give the best impression and seem to have the most up-to-date approach and techniques for maintenance.

Vocabulary-building activities

1. 1 check, inspect, monitor
2 carry out, do, perform
3 flaws, damage, defects
4 collect, gather, obtain
5 remove, repair, replace
6 action, measures, steps

2. 1 preventive maintenance
2 predictive maintenance
3 condition-based maintenance
4 risk-based maintenance
5 reliability-centred maintenance
6 corrective maintenance
7 breakdown maintenance

Unit 7

Focused reading

1. In picture 1, a rig is being assembled onshore before being towed out to sea.
In picture 2, a rig is being erected at sea.
Picture 1 would be most suitable for inclusion in Alan's article.

2. The rig will be dismantled. First, the topside and steel jacket will be removed and towed back to the yard and disposed of. Then, the pipeline will be decommissioned and the debris will be cleared.

Vocabulary-building activities

1.
1 tow, out
2 put off
3 slow down
4 speed up
5 sort out / deal with
6 set up / hook up
7 carry out
8 carry on
9 work out
10 call off

2.

Erection	Dismantling
attached to the deck	abandoned
installed	broken up for scrap
lifted into place	decommissioned
mounted	disposed of or recycled
secured to the seabed	detached from its mountings
	towed back to shore
	unscrewed

Unit 8

Focused reading

1. Students' own answers.

2. Lars lists the following advantages: it is more flexible than conventional subsea systems; the modular systems can be easily replaced; the operations can be carried out remotely; there is a high level of redundancy; the rig is completely reliable; the mud pump system is specially adapted; and it is environmentally friendly.

Vocabulary-building activities

1. 1
a Technology is certainly going to play a crucial role in increasing recovery.
b In the future, offshore platforms could well be run by robots alone.
c It is quite likely that operators will not need to be transported out to the rigs.
d Recovery costs could easily be reduced by up to 50%.
e These platforms might possibly look like hi-tech warehouses.
f It is highly unlikely that supplies of oil and gas will increase.
g We are fairly convinced that seabed rigs will replace topside platforms for subsea drilling.
h The profitability of many oilfields will undoubtedly depend on the prices of oil and gas.
i We are pretty sure that the oil and gas industry will continue to play a major role in the global economy

2 *Confident forecasts:* is certainly going to, could well be, could easily be, highly unlikely, undoubtedly, pretty sure that
Less confident forecasts: quite likely, might possibly, fairly convinced that

3.
1 unmanned
2 remotely
3 sensors
4 signals
5 autonomously
6 actuators
7 grips
8 manipulates
9 redundancy
10 collision
11 interface
12 recognition

Section 3

Unit 1

Focused reading

1.
a 5 inches (127 mm)
b to keep pressure on the bit, to clean, to cool and to lubricate
c they are swept up by the mud and then checked for abnormalities
d 30 ft (10 metres)
e sometime before the 8th century
f from two Greek words meaning 'rock oil'
g 40%

2. 90

Vocabulary-building activities

1.

Hoisting system	Rotating system	Circulating system
derrick crown block travelling block hook swivel	rotary table kelly drillstring drillpipes drill collars bit	mud hose mud pump mud cuttings blowout preventer (BOP) sump pit

2.

Adjective	Noun
long	length
wide	width
deep	depth
high	height
thick	thickness
heavy	weight
old	age
expensive	price

3. 1 a How deep
b What is the pressure
c What is the height
d What is the wind load
e What is the diameter
f How thick
g How long
h What is the cost

2 around, can be as much as, approximately, anything up to, more or less, just under, roughly, ranging from … to …

Unit 2

Focused reading

1. 1 A cantilever mast. The truck has brought the mast to the site and it is about to be erected.

2 a on the ground at the foot of the mast
b at the top of the mast
c somewhere below the crown block

2. The effects of vibration on a drilling mast.

3. 1 It is a sales document for drilling equipment. Its purpose is to convince the purchasing (or 'procurement') manager of the benefits of the NOV's Active Heave Drawworks (AHD) system over those of their competitors.

2 The 'drilling operational window' is increased by allowing drilling programs to continue in heavier seas than conventional drawworks.

Vocabulary-building activities

1. 1 drilling line
2 drawworks
3 crown block
4 travelling block
5 hook

2. Correct order: 3, 1, 5, 2, 4

3. 1 i, 2 f, 3 b, 4 h, 5 a, 6 c, 7 e, 8 g, 9 d

Unit 3

Focused reading

1. To improve rig safety, extend rig operating life, increase drilling efficiency and raise cost-effectiveness. The aim is to achieve performance improvements of between 10 and 30%.

2. The drive system's induction motors are robust and virtually maintenance-free.
It has reduced weight and saves space.
The motor is versatile, reliable and economical.

Vocabulary-building activities

1. 1 transformer
2 engine / generators
3 power management
4 converter panels
5 switchboards
6 console
7 top drive

2. Suggested answers:
1 lower
2 more profitable
3 greater
4 more compact, less
5 more, less
6 simpler, more efficient
7 more reliable, longer

Unit 4

Focused reading

1. 1 Slide 1 – a
2 Slide 2 – e
3 Slide 3 – d
4 Slide 4 – b
5 Slide 5 – c

Vocabulary-building activities

1. 1 tool joints
2 riser
3 casing
4 bit
5 drill collar
6 stabilizers
7 drillstring
8 joint
9 stand
10 tricone bit
11 tubular
12 subs

2. 1 screws
2 flows
3 rotate (or turn)
4 crush
5 drills
6 turns (or rotates)
7 creates
8 wears out
9 decreases, makes

3. a rotary table
b master bushing
c kelly bushing
d kelly
e swivel
f hook

Unit 5

Focused reading

1. 1 lift and carry cuttings
2 allow cuttings to drop
3 prevent cuttings from settling
4 create a film (to prevent caving)
5 seal the borehole
6 keep (underground) pressures in check
7 cool and clean
8 lubricate

2. Water-based fluids are for drilling in environmentally sensitive areas where emulsion fluids cannot be used. They must 'emulate' emulsion systems.

Specialty fluids have additives which enable them to perform specific functions.

Emulsion fluids are more effective and can be used in HPHT situations as long as they are compliant with environmental laws.

Vocabulary-building activities

1. 1 a surface casing
b nozzles
c annulus
d shale shaker
e mud pit

2 a True
b True
c False. They are used to separate cuttings from the mud.
d True
e False. It is made safe and then recirculated.
f True

2.

Functions	Properties	Ingredients
improvement of ROP	density and gel strength	clays, such as bentonite
lubrication of drilling tools	lubricity and viscosity	detergents and thinners to improve efficiency
maintenance of pressure	solubility in water	various polymers for specific purposes
protection of the borehole	toxicity and danger to the environment	water or a synthetic alternative
reduction of fluid loss during drilling		
suspension of cuttings		

3. Possible answers:
1 ... clays, such as bentonite, and water or a synthetic alternative.

2 ... improvement of ROP and the protection of the borehole.
3 ... density and gel strength, and lubricity and viscosity.

Unit 6

Focused reading

1. The problem is that there is a fish in the well. They have hired a fishing services company to retrieve it.
2. Blowout (solutions not discussed)
   Fish (solutions not discussed)
   Stuck pipe (solutions: using a jar or a reamer)
   Lost circulation or sloughing (solutions: use thinner mud, thicker mud, additives, etc.)

Vocabulary-building activities

1.
1 falls into
2 adhere to / pass through
3 got rid of
4 forced into
5 escapes from
6 lead to
7 forces, up

2.

Phrasal verb	Meaning
go on	happen / take place
trip out	pull the drillstring out of the hole
bring in	hire / recruit / use
turn up	arrive
set up	prepare everything to start work
go up	increase
do about	do something in order to solve a problem
thin down	make less thick by adding extra water or other fluid

3. Preventions: 1, 4, 6, 7
Solutions: 2, 3, 5, 8, 9

Unit 7

Focused reading

1. Correct order: 3, 5, 2, 1, 4
2. In the first incident, the pressure and the flow continued to rise. The rig workers were unable to regain control, but were evacuated before fire broke out. In the second incident, the immediate damage appears not to have been as great, but there were a number of injuries because the rig had not been evacuated in time. At the time of the conversation, well control had not yet been called and the flow was not yet under control.

Vocabulary-building activities

1.
1 abandoned
2 killed
3 circulated
4 evacuated
5 destroyed
6 closed
7 broken
8 made
9 ignited
10 called in

2.
1 after
2 during
3 after
4 during
5 before
6 during

3.
1 pipe seal rams
2 blind rams
3 shear rams
4 slip seal rams

Unit 8

Focused reading

1.
1 Extract 1: Directional drilling (and/or slant drilling, deviation drilling) and horizontal drilling
Extract 2: Multilateral drilling
Extract 3: Coiled tubing drilling
Extract 4: Slimhole drilling
Extract 5: Slimhole and coiled tubing drilling

2 a – Extract 5
b – Extract 4
c – Extract 1
d – Extract 3
e – Extract 2

Vocabulary-building activities

1. 1 conventional drilling
2 slimhole drilling
3 coiled tubing drilling
4 slant drilling
5 horizontal drilling
6 multilateral drilling

2. 1 better for the environment / more easily avoided
2 four times greater
3 more economical / easier to transport
4 far less noise / much better insulated
5 more effectively / fewer cuttings
6 far quieter / less disruptive

3. 1 In conventional drilling, formation fluids are prevented from entering the well as far as possible. **However**, in UBD, formation fluids are allowed to flow into the well.
2 UBD is quite an expensive method of drilling. **Nevertheless**, it is widely used because it reduces formation damage.
3 In UBD, the pressure in the well is kept lower than the pressure of the formation. **Therefore,** fluids are forced to flow into the well.
4 In UBD, an inert gas is injected into the drilling mud to reduce its density and hydrostatic force throughout the well depth. **Thus,** the pressure of the well is reduced to below that of the formation.
5 In conventional drilling, formation fluids are prevented from entering the well as far as possible, **whereas** in UBD, formation fluids are allowed to flow into the well.
6 UBD is quite an expensive method of drilling, **yet** it is widely used because it reduces formation damage.
7 In UBD, the pressure in the well is kept lower than the pressure of the formation, **so that** fluids are forced to flow into the well.

Section 4

Unit 1

Focused reading

1. An economic reservoir will normally show the following: Porosity/permeability – 18% for oil, 12% for gas; water saturation – less than 60%.

2.

	Source, reservoir rock or both?	Typical colour(s)	Typical location(s)
Sandstone	Reservoir	White or buff or dark	Beaches, river channels, dunes
Shale	Both	Green or grey or black	River flood plains, ocean bottoms, lakes, lagoons
Limestone	Reservoir (sometimes source)	White or grey	Desert areas, e.g., Middle East

Vocabulary-building activities

1. 1 source
2 reservoir, reservoirs
3 seal, seals
4 trap
5 maturation
6 migration

2.

Verb	Adjective	Noun
	sedimentary	sediment
	porous	**porosity**
permeate	**permeable**	**permeability**
migrate		migration
saturate	saturated	**saturation**
accumulate		**accumulation**
fracture	**fractured**	fracturing
expel	**expelled**	**expulsion**

3. 1 a sand dunes
b lagoon

c delta

2 Onshore: coastal plain, shorelines, dunes
Offshore: trench, ridge, delta, continental shelf, continental slope, canyon
Either: basin, reefs / lagoon

Unit 2

Focused reading

1. 1 The geophysicist describes Prudhoe Bay's geology because Carson Sound is nearby and has a similar structure.

2 The field is sealed by various types of traps: to the north, a normal fault; to the south, south-dipping shales; to the west, a saddle; to the east, unconformities and shales.

2. 1 Mukluk was another field near Prudhoe Bay with a similar geology.

2 The manager should perhaps be worried because Mukluk was a financial disaster.

Vocabulary-building activities

1. 1 salt dome
2 truncation
3 fault trap
4 pinch-out
5 anticline

2. Correct order: 4, 2, 3, 1, 5

Unit 3

Focused reading

1. To make contact with the potential customer and ask for a meeting to present the consultant's findings in West Africa.

2. 1 Identification of farm-in opportunities
2 Reserves assessment
3 Studies of individual fields

Vocabulary-building activities

1. 1 Studies of individual fields
2 Risk assessment and chance of success
3 Regional evaluations
4 Open-acreage assessment
5 Reserves assessment

2. 1 would / could
2 should
3 would
4 may
5 would
6 could / would
7 would
8 would
9 may

3. 1 a) Thanks b) call / come / drop
2 c) Let d) pleased
3 e) back f) grateful / pleased / happy
4 g) wish
5 h) Do / Please i) hesitate j) further / other / more
6 k) Looking l) earliest

4. 1 Informal: Hi James, Dear Olga, Hello Pete
Formal: Dear Ms O'Hea, Dear Sir

2 Informal: Best wishes, Kind regards, Take care
These can be used with all three informal openings, but it will depend on how close the relationship is. 'Take care' is more friendly. The other two are more business-like.
Formal: Yours sincerely, (follows Ms O'Hea), Yours faithfully (follows Dear Sirs)

Unit 4

Focused reading

1. Students' own answers.

Vocabulary-building activities

1. Isopach map

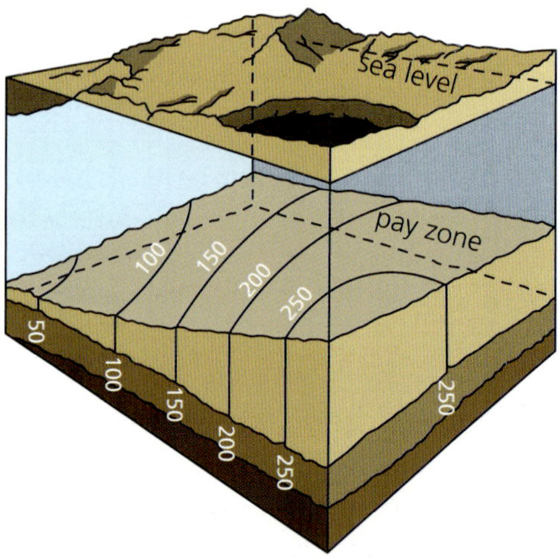

Topographic map

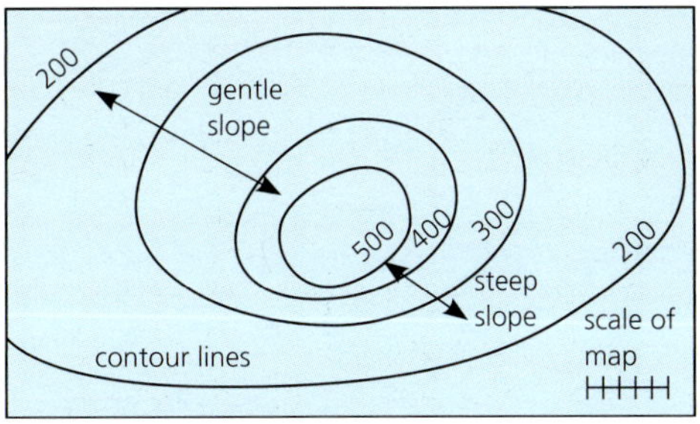

Geologic map

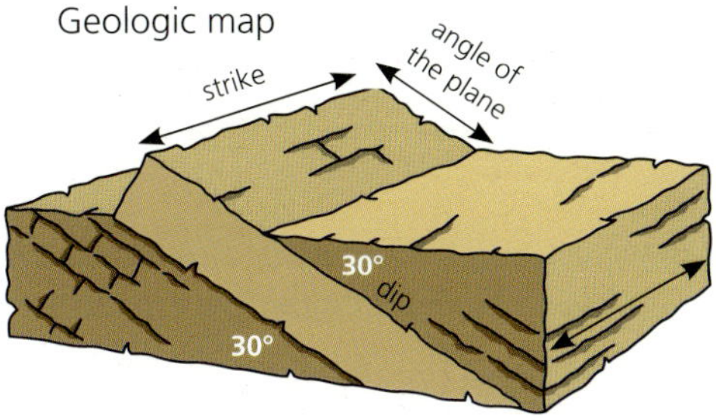

2. 1 1 Can you tell us what a reservoir model consists of?
2 How can seismic data be incorporated into cross sections?
3 Do you know what SIGMA stands for?
4 Can you explain how old maps are integrated with new ones?
5 How do they capture field data?
6 Sorry, but I don't quite understand what you mean by GIS software.
7 Can you explain what an isopach map is?
8 What does a net pay map show?
2 Students' own answers.

Unit 5

Focused reading

1. 1 By listening for echoes through geophones. The acoustic contrast between a reservoir of gas and the cap rock that seals it is significant because gas slows down the sound waves.
2 It is more difficult to detect oil because the contrast between oil and the surrounding rock is less significant.

2. 1 By reducing or removing random noise.
2 4D seismic is done over a period of time, maybe even years. Time is the extra dimension.

Vocabulary-building activities

1. Sentences referring to the *source*: 1, 5, 6, 7, 8, 9, 11 and 12.
Sentences referring to the *receiver*: 2, 3, 4, 10 and 13.

2. Correct sentences are as follows:
1 The greater …, the longer the travel time …
2 … is calculated by multiplying sound velocity by density …
3 The greater the contrast …
4 … have a very low reflection coefficient.
5 Gas reduces the velocity of sound …
6 It is more difficult to tell the difference …
7 … where the contrast in impedance is very high.
8 … a petroleum trap in a layer of rock which is not flat.

Unit 6

Focused reading

1. The following can be labelled with the information provided:
Black line: ROP (rate of penetration), feet per hour
Blue line: porosity
Yellow squares: sandstone
Grey/black squares: shale
Green line: gas (parts per million)

2. Key points to be underlined are:
… Depth, Rate of Penetration, Total Gas and Chromatograph component gases …
… accurate depiction of Show Zone character …
… a tremendous database to utilize Pixler Ratios to evaluate the wellbore.

Vocabulary-building activities

1. 1 trailer
2 manned
3 tours
4 submitted
5 interpretations
6 cuttings
7 reports

8 instrumentation
9 remotely
10 compatibility
11 loggers
12 zone of interest

2. 1 lithologic log
2 depth strip
3 rock sample
4 core
5 sidewall coring
6 wireline well log
7 sonde
8 openhole log
9 cased-hole logging

3. assessing, calculating, checking, extrapolating, identifying, mapping, measuring, monitoring, plotting, sampling

Unit 7

Focused reading

1. 1 A sonde is a cylinder containing wireline logging tools.
2 It is lowered to the bottom of the well and then run up again to allow the tools to acquire and transmit data.

2. Students' own notes.

Vocabulary-building activities

1. A armoured cable
B expandable arms
C bow-springs

2.

Name and abbreviation	Alternative name	Track	Use
Spontaneous Potential (SP)	Self Potential	1	For identifying rock formation using specific SP signature
Induction Log (IL)	No information	2	For measuring resistivity
Gamma Ray Log (GRL)	No information	1	For measuring radioactivity
Neutron Porosity Log (NL)	No information	2	For measuring porosity
Formation Density Log (FDL)	Gamma-Gamma Log	2	For measuring density
Caliper Log (CAL)	No information	1	For measuring diameter of well

3. 1 c, 2 f, 3 a, 4 e, 5 d, 6 b

Unit 8

Focused reading

1. 1 The oilfield shows a mix of rock from the Carboniferous (the fault) and Jurassic periods (source and reservoir).
2 The trap was created by an anticline above a basement fault, with an evaporitic package of rocks as the seal.
3 The reservoir rocks are made of limestone, with very high porosity. They are sourced from a marine shelf deposit of mud and lime with very high organic material content.

2. 1 The oil was formed 500 million years ago. The rock is pre-Cambrian and has very high salinity.
2 The rock is so ancient, the temperature of the formation is extremely low and the geological structure is very complex.

Vocabulary-building activities

1. 1 gives fuller explanation of the phrase 'as the north-east margin of Africa began to impinge on south-west Asia …'
2 gives fuller explanation of the phrase 'The seal is an evaporitic package of rocks …'
3 gives fuller explanation of the phrase 'previously inaccessible crude oil reserves …'
4 gives fuller explanation of the phrase 'there is almost no experience in the industry in dealing with such ancient pre-Cambrian rock …'
5 gives fuller explanation of the phrase 'the company is experimenting with ways to reduce the salinity …'
6 gives fuller explanation of the phrase 'They

gave up because of the difficulties and lack of infrastructure.'

2. 1 T**R**AP
2 SC**A**LE
3 MIGRA**T**E
4 G**E**OPHONE
5 S**O**URCE
6 **F**OLD
7 **P**LANE
8 SHAL**E**
9 SO**N**DE
10 D**E**PLETION
11 SANDS**T**ONE
12 PO**R**OUS
13 SE**A**L
14 S**T**EEP
15 L**I**THOLOGY
16 VEL**O**CITY
17 CURRE**N**T

The key measurement for drillers is the RATE OF PENETRATION.

Section 5

Unit 1

Focused reading

1. a complete line
high-performance
proven to work in the most demanding environments
specifically designed to handle the increased demands
designed for maximum erosion resistance
positive sealing integrity
a reliable one-trip perforating and Gravel Pack/Frac Pack system

2. 1 In cased-hole completion, there is casing all the way to the bottom of the hole. Perforations are made through the casing to allow fluids to flow into the well. In openhole completion, there is no casing in the last reservoir section and the formation is held in place by other means, such as screens, gravel packs, etc.
2 The flow can be improved by stimulation with acid or by fracturing the formation.
3 Methods mentioned include: installing a small-diameter tubing, various pumps and jacks gas lift, and all-in-one completion and improvement systems (multiple packer systems).

Vocabulary-building activities

1. Preparing the well: 12, 6, 15
Running casing: 4, 7, 10, 11
Cementing: 3, 5, 1, 9, 14, 8, 13, 2

2. Openhole completions: 1, 3, 4, 6, 7, 8
Cased-hole completions: 2, 5, 9

3. a Christmas tree
b Wellhead
c Casinghead
d Tubing head
e master valve
f wing valve
g pressure gauge

Unit 2

Focused reading

1. Onshore separators can be categorized as horizontal or vertical, two-phase or three-phase, and according to their operating pressure.
The most important force for separation is gravity. Other factors involved are operating pressure, residence time and flow type.

2. Primary separation occurs when centrifugal force separates the heavier liquids and they fall to the bottom of the vessel.
Secondary separation occurs in the vortex finder tube, with spinning taking place at a much greater velocity.

Vocabulary-building activities

1. 1 b, 2 c, 3 a, 4 c, 5 d, 6 d, 7 a, 8 b

2. 1 conditioning
2 stripping
3 dehydration
4 sweetening
5 compression
6 cooling
7 absorption

3. • *Stock tanks* are for storing oil.
   • *A gauge tape* is for measuring the amount of oil in the stock tank.
   • *A shake-out test* is for measuring the amount of impurities in the oil.
   • A *turbine meter* is for measuring the volume of flowing oil.
   • *Charcoal tests* are for measuring the condensate content of gas.
   • *Fractional analysis* is for measuring the relative amounts of hydrocarbon and water in gas.
   • *An orifice meter* is for measuring the volume of flowing gas.

Unit 3

Focused reading

1. 1 Picture A is a drilling rig, picture B is a workover rig.
   2 The workover rig has been hoisted from a truck, whereas the drilling rig is fixed in position. A workover rig is always a temporary construction.

2. 1 a sand-control job, a cement squeeze job, a fishing job and sidetracking
   2 A *sand-control job* is necessary when there is a build-up of sand in the wellbore.
   A *cement squeeze job* is necessary when there are gaps in the cement behind the casing.
   A *fishing job* is necessary when a piece of equipment falls downhole.
   Sidetracking is necessary when you need to avoid a blockage in the well.

Vocabulary-building activities

1. 1e, 2 g, 3 f, 4 c, 5 a, 6 d, 7 h, 8 b

2. 1 removal
   2 completed
   3 restore
   4 lower
   5 wireline
   6 hollow
   7 raised
   8 piston
   9 attached
   10 pressure
   11 killed
   12 retain

3. 1 excess material
   2 collapsed casing, obstacles
   3 scale and debris

4. 1 during, on, above, by
   2 For, to, to, in
   3 In, around, from, into
   4 by, with, down, under, until

Unit 4

Focused reading

1. well stimulation
   use of pumps
   sidetrack drilling

2. well stimulation
   multiple fracturing
   acidizing
   enhanced production

Vocabulary-building activities

1. Correct order: 5, 4, 6, 3, 7, 1, 2

2. 1 … when put into contact with suitable reservoir rocks, …
   2 … known as acidizing.
   3 … high enough to cause it to fracture.
   4 … which is used mainly in limestone formations …
   5 … keeping the injection pressure low enough to avoid fracturing the formation.
   6 … that block reservoir permeability …
   7 … a strong acid that works best on carbonate rocks.
   8 … that have high temperatures –250 degrees Fahrenheit or more.
   9 … which reacts with quartz, sand and clay.
   10 Regardless of the acid used …

3. cut, enhanced, improved, increased, reduced

Unit 5

Focused reading

1. 1 Primary recovery is the recovery achieved by the natural drive of the reservoir. Secondary recovery is when water is injected into the well to force the oil towards the producing wells.

Tertiary recovery, also called enhanced recovery, uses additional techniques to enhance primary or secondary recovery.
   2 US producers are using tertiary recovery because most easy-to-produce oil has already been recovered.

2. Example notes:
   1 *Thermal recovery* uses heat, for example, in the form of steam, to lower viscosity and improve flow. Fifty per cent of US EOR uses thermal recovery methods.
   2 *Gas injection* works in two different ways: either a) by using gases that expand and force oil towards producing well; or b) using gases that dissolve in the oil and lower viscosity. Fifty per cent of US EOR uses one or another of these gas injection methods.
   3 *Chemical injection* lowers the surface tension to increase flow of oil. It accounts for less than 1% of US EOR.

3. Example notes:
   1 The *Dakota Gasification Company project* is producing gas and delivering it via a 204-mile pipeline to Encana's oilfield in Canada. By injecting CO_2, Encana hopes to add 25 years and 130 million barrels of oil to its planned recovery.
   2 The *Hall-Gurney* project is trialing the use of CO_2 injection to gain energy, environmental and economic benefits.
   3 A *second Hall-Gurney* project involves testing the feasibility of 4D seismic (time-lapse) monitoring of CO_2 injection to improve efficiency and economy.

Vocabulary-building activities

1. 1 Dissolved Gas Drive
   2 Free Gas Cap Expansion Drive
   3 Water Drive

2. 1 The water forces some of the remaining oil through to producing wells.
   2 … injecting a gas that dissolves in the oil (usually carbon dioxide or liquid petroleum gas) into the reservoir.
   3 … injecting different chemicals, in separate slugs, each serving a different purpose, …
   4 Steam is injected into the reservoir to heat the oil and make it more fluid.
   5 … air is injected into the reservoir to sustain the fire.

Unit 6

Focused reading

1. The system is dual purpose because it detects both leaks and breaches of security. It uses a single fibre-optic cable run adjacent to the pipeline.

2. 1
   - Early warning of leaks
   - Early warning of potential disruption
   - Immediate and accurate location of disturbance
   - Easy to install and low maintenance

   2
   - Immunity to interference
   - Function in all weather and soil conditions

3. Students' own notes.

Vocabulary-building activities

1. Example notes:
   Initial Injection Station: product injected into line
   Partial Delivery Stations: allows partial product delivery
   Compressor/Pump Stations: moves product through pipeline
   Block Valve Stations: isolates segment of line
   Final Delivery Station: product distributed to consumer

2. 1 a 6-mile stretch
   2 10 to 120 centimetres
   3 of about 1 to 2 metres
   4 5 and 20 feet per second
   5 2 inches to over 60 inches
   6 12 and 24 metres long

3. 1 i, 2 g, 3 d, 4 c, 5 b, 6 a, 7 e, 8 f, 9 h

Unit 7

Focused reading

1. 1 Extract 1: a specialist oil and gas industry magazine.
   Extract 2: a daily newspaper.
   Extract 3: an online encyclopedia.
   2 This answer will depend on factors such as the country, the type of newspaper, etc.,

but most likely answers might be:
Extract 1: Business
Extract 2: World News or Environment
Extract 3: Business or Environment

Vocabulary-building activities

1. Example answers:
1 We've increased our profits five times!
2 They've made ten times more than this time last year.
3 First, there aren't enough tankers, and second, OPEC has increased its production.
4 $180,000 per day.
5 $17,000 per day.

2. Example answers:
1 Just off the south coast.
2 There were heavy storms.
3 Yes, a slick has washed ashore.
4 67,500 tonnes
5 No, there are still 40,000 tonnes on board.
6 No, the seafront has been closed.
7 Yes, it has polluted the beaches and hundreds of fish, birds and turtles are dead.
8 A single-hulled tanker.

3. 1

Source of oil spill	**% by volume of oil spilt**	**% by number of incidents**
Oil tankers	35.7	5
Facilities, etc.	27.6	not known
Non-tank vessels	19.9	51.8
Pipelines	9.3	not known
Mystery spills	7.4	not known

2 Percentage of spills from routine operations, under seven metric tons per spill = <u>91%</u>
Percentage of spills from accidents, over 700 metric tons per spill = <u>84%</u>

3 Model answer:
Most oil by volume is spilt from oil tankers, but with relatively few incidents. Accidental spills account for a very large number of large spills. Spills from routine operations are small but frequent.

Unit 8

Focused reading

1. The company performs full crude oil assay assessments and detailed hydrocarbons analyses, including: petroleum assay testing and analysis, reporting, crude oil characterization, methanol and mercury monitoring and analysis, assay data entry and graphic display, database storage of results, comparison of current and historical assay data.

2. *Premium* refers to the quality – above 30 API.
Light refers to the density of the crude oil – low density.
Sweet refers to the amount of sulphur – low quantities of sulphur.

Vocabulary-building activities

1. 1 proportions
2 whereas
3 types
4 determined
5 dependent
6 content
7 yield
8 subjected
9 form
10 liquefied

2. 1 Hungo
2 Grane, with a density of below API 20.
3 Azeri Light and Ekofisk.
4 Hungo Blend, with medium sulphur content.
5 Lufeng is the only one which is waxy.
6 Azeri Light, Lupeng and Ekofisk.

3. Students' own answers.